Chaotic dynamics
an introduction

Chaotic dynamics
an introduction

GREGORY L. BAKER

Academy of the New Church College
Bryn Athyn, Pennsylvania

AND

JERRY P. GOLLUB

Haverford College, Haverford, Pennsylvania

CAMBRIDGE
UNIVERSITY PRESS

Published by the Press Syndicate of the University of Cambridge
The Pitt Building, Trumpington Street, Cambridge CB2 1RP
40 West 20th Street, New York NY 10011, USA
10 Stamford Road, Oakleigh, Melbourne 3166, Australia

First published 1990
Reprinted 1990 (with corrections), 1991 (twice), 1992

Printed in Canada

British Library cataloguing in publication data

Baker, Gregory L.
Chaotic dynamics.
1. Nonlinear dynamical systems. Chaotic behaviour
I. Title   II. Gollub, Jerry P.
515.3'5

Library of Congress cataloging in publication data

Baker, Gregory L.
Chaotic dynamics : an introduction / by Gregory L. Baker and Jerry
P. Gollub.
p.   cm.
ISBN 0-521-38258-0   ISBN 0-521-38897-X (pbk.)
1. Pendulum.   2. Chaotic behavior in systems.   I. Title.
QA862.P4B35   1990
003-dc20   89-22311   CIP

ISBN 0 521 38258 0 hardback
ISBN 0 521 38897 X paperback

Contents

Preface

The remarkable fact that determinism does not imply either regular behavior or predictability is having a major impact on many fields of science, engineering, and mathematics. The discovery of chaos changes our understanding of the foundations of physics, and has many practical applications as well. The subject sheds new light on the workings of lasers, fluids, mechanical structures, and chemical reactions.

Interest in chaos (or more generally, nonlinear dynamics) has grown rapidly since 1963, when Lorenz published his numerical work on a simplified model of convection, and discussed its implications for weather prediction. The research literature has exploded, and many books on chaotic dynamics have appeared. Yet most of these works are directed at specialists or graduate students. These works include, for example, the very challenging but authoritative work *Nonlinear oscillations, dynamical systems, and bifurcations of vector fields*, by J. Guckenheimer and P. Holmes (Springer Verlag NY, 1983) and *Chaotic vibrations*, by F.C. Moon (John Wiley and Sons, NY, 1987).

At the other end of the spectrum of accessibility, a few popular books and articles give the flavor of chaos but do not allow the reader a significant measure of participation in the subject. Examples of this latter category include the *Scientific American* article by Crutchfield *et al.*, entitled 'Chaos' in the December issue of 1986 (pp. 46–57), and the very enjoyable book by James Gleick, *Chaos: making a new science* (Viking, NY, 1987).

The present work fills part of the gap left between these two types of publications. *Chaotic dynamics: an introduction* introduces chaotic dynamics through the study of the driven pendulum, a simple system whose nonlinear properties are often ignored in teaching mathematics and physics. It is written at a level accessible to sophomore/junior level undergraduate students of mathematics or physics. The work should also be quite readable for secondary and college teachers, or anyone with a few courses in college mathematics and physics. Unlike the popularizations, this work is intended to help the reader develop an understanding of chaotic dynamics and to provide some of the enjoyment of participation through analytic and computer exercises.

Chaotic dynamics: an introduction is written as a short text or text supplement. The following background is assumed: elementary multivariable calculus, linear differential equations, and introductory physics. We have included a heuristic treatment of Fourier analysis. Since many of the exercises are numerical, some programming experience is desirable.

We include listings of useful programs in a widely available form of BASIC known as 'True BASIC^{tm}.' A menu-driven runtime package is available on 5¼ in. diskette at a moderate cost. This requires the following hardware: IBM PC, AT, PS/2, or compatible (such as AT&T 6300), with 512K of memory; and a CGA, EGA, or Hercules graphics adaptor. A math coprocessor is highly desirable. To modify the listed programs, the True BASIC Language System, version 2.0 or later, and the True BASIC Scientific Graphics Toolkit are also required. These are available from True BASIC Inc., 12 Commerce Ave., West Lebanon, NH 03784, USA. (A Macintosh version of the diskette may be available. See diskette order information.)

Chaotic dynamics: an introduction can be used as a textbook for a short course on chaos, or as a supplement for courses on classical mechanics or modern physics. Some of the material can be included in an introductory physics course.

Since this book is not a research review, we have emphasized accessibility rather than completeness. For example, the discussion of a very broad class of models known as Hamiltonian dynamical systems is beyond the level of this book. Such models are required for energy conserving systems and are discussed, for example, in *Non-linear physics: from the pendulum to turbulence and chaos* by R.Z. Sagdeev, D.A. Usikov, and G.M. Zaslavsky, Harwood Academic Publishers, Chur, Switzerland, 1988. References that we judged to provide useful additional information are cited, but they are not intended to provide a balanced assessment of the scholarship of the many scientists and mathematicians who have contributed to this subject.

Gregory L. Baker
Bryn Athyn, Pennsylvania

Jerry P. Gollub
Haverford, Pennsylvania

Acknowledgments

We are pleased to acknowledge the support and encouragement of many people during the development of this book.

One of us (GLB) is grateful to the administration and directors of the College of the Academy of the New Church for a sabbatical leave in 1988, when a major portion of the work was done. During this period many of his responsibilities were graciously assumed by colleagues; notably Robert Gladish, Grant Doering, Charles Ebert, Michael Brown, and Daniel Goodenough. Other colleagues, too numerous to name, showed support through continued interest and encouragement. One of us (JPG) acknowledges financial support from the National Science Foundation Low Temperature Physics Program and from the Applied and Computational Mathematics (URI) Program of the Defense Advanced Research Projects Agency.

We are grateful to Lelia Howard and Charles Ebert for computer support, Lyle Roeloffs for providing some software, Jeff Gerecht and Miguel Rubio for their scrutiny of the manuscript, and Douglas Withers for helpful conversations. We wish to thank Harry Swinney for permission to reprint the phase portrait shown in Figure 6.6, John Wiley and Sons for permission to reprint the phase portrait in Figure 6.2, Alain Arneodo for his communication in regard to dimension calculations, and Robert Hellemen for his interest during the early stage of the project. We appreciate extensive help provided by librarians Doreen Carey, Carroll Odhner, and Suzanne Newhall at the Academy and Haverford College.

Finally, we thank our wives Margaret Baker and Diane Nissen for their constant support and for tolerating some neglect of family responsibilities.

CHAPTER ONE

Introduction

The irregular and unpredictable time evolution of many nonlinear systems has been dubbed 'chaos.' It occurs in mechanical oscillators such as pendula or vibrating objects, in rotating or heated fluids, in laser cavities, and in some chemical reactions. Its central characteristic is that the system does not repeat its past behavior (even approximately). Periodic and chaotic behavior are contrasted in Figure 1.1. Yet, despite their lack of regularity, chaotic dynamical systems follow deterministic equations such as those derived from Newton's second law.

The unique character of chaotic dynamics may be seen most clearly by imagining the system to be started twice, but from slightly different initial conditions. We can think of this small initial difference as resulting from measurement error, for example. For nonchaotic systems this uncertainty leads only to an error in prediction that grows *linearly* with time. For chaotic systems, on the other hand, the error grows *exponentially* in time, so that the state of the system is essentially unknown after a very short time. This phenomenon, which occurs only when the governing equations are nonlinear, is known as *sensitivity to initial conditions*. Henri Poincaré (1854–1912), a prominent mathematician and theoretical astronomer who studied dynamical systems, was the first to recognize this phenomenon. He described it as follows: '. . . it may happen that small differences in the initial conditions produce very great ones in the final phenomena. A small error in the former will produce an enormous error in the latter. Prediction becomes impossible, and we have the fortuitous phenomenon' (Poincaré, 1913).

If prediction becomes impossible, it is evident that a chaotic system

1

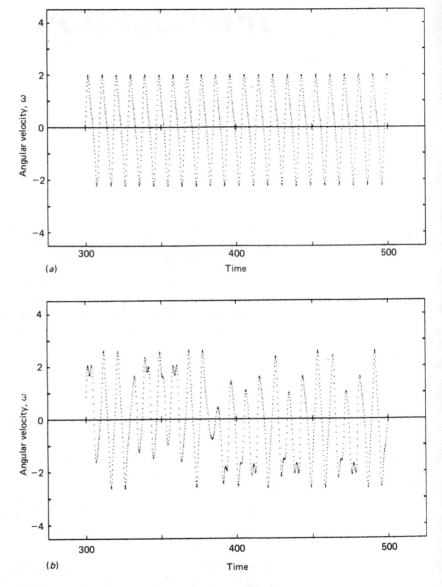

Fig. 1.1 The damped,
driven pendulum can
exhibit both periodic and
chaotic motions. Here,
the angular velocity is
shown as a function of
time for the two cases.

can resemble a stochastic system (a system subject to random external forces). However the source of the irregularity is quite different. For chaos, the irregularity is part of the intrinsic dynamics of the system, not unpredictable outside influences.

Chaotic motion is not a rare phenomenon. Consider a dynamical system described by a set of first order differential equations. Several necessary conditions for chaotic motion are that (*a*) the system has at least three independent dynamical variables, and (*b*) the equations of motion contain a nonlinear term, that couples several of the variables. The equations can often be expressed in the form:

$$dx_1/dt = F_1(x_1, x_2, \cdots, x_n)$$
$$dx_2/dt = F_2(x_1, x_2, \cdots, x_n)$$
$$\vdots$$
$$dx_n/dt = F_n(x_1, x_2, \cdots, x_n)$$

where *n* must be at least 3. Two examples of appropriate nonlinear equations are:

$$dx_1/dt = \alpha x_1 + \beta x_2 + \gamma x_1 x_2 + \cdots + \delta x_n$$
$$dx_1/dt = \alpha x_1 + \beta x_2 + \gamma \sin x_2 + \cdots + \delta x_n$$

where α, β, γ, δ are constants. In each case the nonlinear term couples both x_1 and x_2. Systems such as these are often chaotic for some choices of the constants.

The fact that only three variables are required for chaos was surprising when first discovered. We shall see that three-space is sufficient to allow for (*a*) divergence of trajectories, (*b*) confinement of the motion to a finite region of the phase space of the dynamical variables, and (*c*) uniqueness of the trajectory. The nonlinearity condition is perhaps less surprising. Solutions to linear differential equations can always be expressed as a linear superposition of periodic functions, once initial transients have decayed. The effect of a nonlinear term is often to render a periodic solution unstable for certain parameter choices. While these conditions do not guarantee chaos, they do make its existence possible.

The nonlinearity condition has probably been responsible for the late historical development of the study of chaotic systems. Despite the fact that chaotic systems are deterministic and are described by many of the long-known classical equations of physics, the development of the subject itself is more recent. This circumstance may arise from the fact that, with the exception of some first order equations, nonlinear differential equations are either difficult or impossible to solve analytically. Although it is sometimes possible to use linearized approximations, the solution of nonlinear differential equations generally requires numerical methods whose practical implementation demands the use of a digital computer. The first numerical study to detect chaos in a nonlinear dynamical system was that of Lorenz's model of convective fluid flow (Lorenz, 1963). Similarly, the majority of the diagrams in this book are based upon the use of numerical methods on a personal computer to solve nonlinear equations.

From these general comments on chaotic systems, we turn to the physical system that is the focus of this work – the damped, driven pendulum. The choice of the pendulum as a model system has strong historical precedent in physics. Galileo postulated the constancy of period for small amplitude oscillations of the pendulum from observations of swaying lamps in the cathedral at Pisa, in 1581 (Robinson, 1921). He took up the problem of the relationship between the period and pendulum length in his famous *Dialogue on the Two Principal World Systems* in 1632, and in 1637 he suggested that the square of the period was proportional to the length of the pendulum for small oscillation amplitudes (Dugas, 1958). The pendulum also served as a primary timing mechanism for clocks and as a method of measuring variations in the earth's gravitational field. As a pedagogical device the pendulum has long been a standard mechanical example in introductory physics and classical mechanics courses. Now, 400 years after Galileo's initial work, the pendulum has again become an object of research as a chaotic system. The references scattered throughout this work attest to its popularity.

The damped, sinusoidally driven pendulum of mass m (or weight W) and length l is described by the following equation of motion:

$$ml\frac{\mathrm{d}^2\theta}{\mathrm{d}t^2} + \gamma\frac{\mathrm{d}\theta}{\mathrm{d}t} + W\sin\theta = A\cos(\omega_D t).$$

This equation expresses Newton's second law with the various terms on the left representing acceleration, damping, and gravitation. The angular velocity of the forcing, ω_D, may be different from the natural frequency of the pendulum. In order to minimize the number of adjustable parameters the equation may be rewritten in dimensionless form as:

$$\mathrm{d}^2\theta/\mathrm{d}t^2 + (1/q)\mathrm{d}\theta/\mathrm{d}t + \sin\theta = g\cos(\omega_D t)$$

where q is the damping or quality parameter, g is the forcing amplitude, not to be confused with the gravitational acceleration, and ω_D is the drive frequency. The low-amplitude natural angular frequency of the pendulum is unity, and time is regarded as dimensionless. (This particular notation follows that used by Gwinn and Westervelt. See for example, Gwinn and Westervelt (1986).) This equation satisfies the necessary conditions for chaos when it is written as a set of first order equations:

$$d\omega/dt = -(1/q)\omega - \sin\theta + g\cos\phi$$
$$d\theta/dt = \omega$$
$$d\phi/dt = \omega_D.$$

The variable ϕ is introduced as the phase of the drive term. The necessary three variables (ω, θ, ϕ) are evident, and the $\sin\theta$ and $g\cos\phi$ terms are clearly nonlinear. Whether the motion is chaotic depends upon the values of the parameters g, ω_D, and q. For some values the pendulum locks onto the driving force, oscillating in a periodic motion whose frequency is the driving frequency, possibly with some harmonics or subharmonics. But for other choices of the parameters the pendulum motion is chaotic. One may view the chaos as resulting from a subtle interplay between the tendency of the pendulum to oscillate at its 'natural' frequency and the action of the forcing term. The transitions between nonchaotic and chaotic states, due to changes in the parameters, occur in several ways and depend delicately upon the values of the parameters.

A variety of analytic and computational tools may be used in the study of chaotic systems. In Chapter 2 several of these are discussed. The pendulum's phase space and its properties are described, together with the conceptual device known as the Poincaré section. Then, since Fourier spectra are an indicator of chaotic motion, some elements of Fourier analysis are outlined. Chapter 3 is a description of the application of these and other techniques to the pendulum.

The driven pendulum would seem to be one of the simplest physical systems. Yet its behavior is rich and complex. The study of its motion can be facilitated by simple mathematical models formulated as difference equations, that provide a discrete *mapping* of the system from one state to another. Mappings have the advantage of being conceptually simple and numerically efficient, and they may be used as paradigms for various aspects of the pendulum motion. Chapter 4 contains discussions of three such maps, the logistic map, the circle map, and the horseshoe map. We use them to provide insight into the behavior of the pendulum.

Chapter 5 is concerned with the geometric structure of the *attractor* that describes the chaotic pendulum. The attractor, and its Poincaré section, are *fractal* structures with noninteger dimensionality. Various approaches to the calculation of fractal dimension are described. Another geometric feature is the exponential divergence of the chaotic trajectories on the attractor. The rate of this divergence is characterized by Lyapunov exponents. The calculation of these exponents and their relation to (*a*) the fractal dimension, (*b*) the dissipative nature of the pendulum, and (*c*) the duration of predictable behavior are also discussed.

Finally, in Chapter 6 a few general comments are made on the relationship of chaotic behavior to other areas of physics. While chaotic behavior occurs broadly, three areas are given brief descriptions: fluid dynamics, chemical reactions, and lasers. The relation of chaos to quantum mechanics and the connection of chaos with irreversibility are also discussed briefly.

Two appendices present numerical aspects of this book. Appendix A is a description of the Runge–Kutta algorithm used to solve the pendulum differential equation. Appendix B provides brief descriptions and listings of the computer programs used throughout the text, and in the computer exercises given at the end of several of the chapters. The listings utilize the language True BASIC[tm] but they are adaptable to any compiled BASIC or other high level language. (Interpreted BASIC, which is typically delivered with current microcomputers, is too slow for most of these simulations. The exceptions are the mappings in Chapter 4.)

CHAPTER TWO

Some helpful tools

In this chapter we discuss three mathematical constructs that are generally useful in the study of dynamical systems: phase space, the Poincaré section, and power spectra. Phase space is the mathematical space of the dynamical variables of a system. The Poincaré section is a 'snapshot' of the motion in the phase space, taken at regular time intervals. The power spectrum is computed using Fourier analysis to display the frequency composition of the time variation of the dynamical variables.

Phase space

The phase space of a dynamical system is a mathematical space with orthogonal coordinate directions representing each of the variables needed to specify the instantaneous state of the system. For example, the state of a particle moving in one dimension is specified by its position (x) and velocity (v); hence its phase space is a plane. On the other hand, a particle moving in three dimensions would have a six-dimensional phase space with three position and three velocity directions. A phase space may be constructed in several different ways. For example, momenta can be used instead of velocities. ·

Let us focus the discussion on the pendulum and begin with the familiar simple pendulum in the small amplitude approximation where the restoring term, $\sin\theta$, is taken as θ. (Recall that the equations are written in dimensionless form for simplicity, with time measured in units of the inverse of the natural frequency.) The equation of motion is

$$d^2\theta/dt^2 + \theta = 0.$$

With the addition of the angular velocity variable, $\omega \equiv d\theta/dt$, this linear, second order equation can be reduced to two first order equations:

$$d\omega/dt = -\theta$$

and

$$d\theta/dt = \omega.$$

In this way each dynamical variable has its own first order differential equation. Without loss of generality, the initial conditions can be chosen so that the solution becomes

$$\theta = a_i \cos t \text{ and } \omega = a_i \sin t$$

where $\{a_i\}$ represents the possible amplitudes of the motion. This solution set gives the parametric curves for ω and θ, and one can eliminate the time parameter to give a two-dimensional representation for differing values of a_i. This diagram, shown in Figure 2.1, is

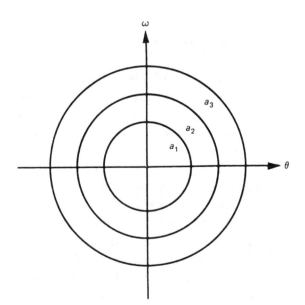

Fig. 2.1 Phase diagram of the linear pendulum. The angular velocity, ω, and the angular displacement, θ, are the coordinate axes.

the appropriate *phase space diagram* (in this case a phase plane diagram). Each value of a_i yields a closed orbit of fixed energy. The energy increases with the square of the radius a_i. The orbit is usually called a *phase trajectory*.

An important feature of the trajectory is that two trajectories corresponding to similar energies will pass very close to each other, but the orbits will not cross each other. This *noncrossing* property derives from the fact that past and future states of a deterministic mechanical system are uniquely prescribed by the system state at a given time. A crossing of trajectories at time t would introduce ambiguity into past and future states, thereby rendering the system indeterminate. Such indeterminacy would contradict the assumed uniqueness of the trajectory. Figure 2.2 shows the indeterminacy of trajectories emanating from a hypothetical crossing.

Another important feature of the phase space of *conservative* (constant energy) systems is the *preservation of areas*. This means that all the points found in a given area of phase space at one time move in such a way that at a later time the area occupied by these points remains the same. This feature is illustrated in Figure 2.3 and in Examples 2.1 and 2.2.

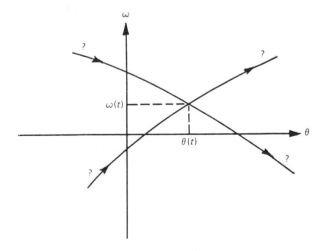

Fig. 2.2 The noncrossing property of phase trajectories. Crossing of trajectories violates uniqueness of trajectories in a deterministic dynamical system.

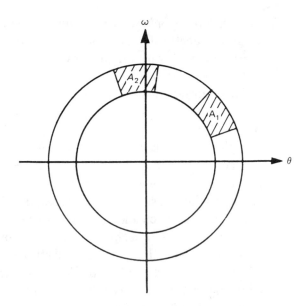

Fig. 2.3 Preservation of
phase space area.

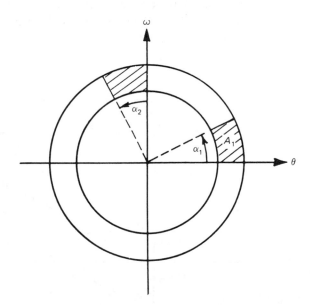

Fig. 2.4 Evolution of the
linear oscillator described
in Example 2.1.

Example 2.1. For the linear oscillator $d^2\theta/dt^2 + \theta = 0$ consider the
evolution of the area A_1 as shown in Figure 2.4 during one quarter of
the period. Since the system is energy conserving, A_1 should remain
constant. Because of the circular symmetry, preservation of the area

can be shown by proving that every point in A_1 rotates (at a constant radius) through the same angle in the quarter period. The energy conserving feature ensures that each point rotates at a constant radius because the energy of the oscillator is proportional to the square of the radius. For the rotation angle we note that since $\theta = a\cos t$ and $\omega = a\sin t$, the polar angle of a given point is

$$\alpha(t) = \tan^{-1}(\tan t) = t.$$

Therefore at $t = t_0 + \pi/2$, $\alpha(t_0 + \pi/2) = t_0 + \pi/2$. But since $\alpha(t)$ was arbitrary, all points rotate by $\pi/2$ in one quarter period, and the area is preserved.

Example 2.2. As another example of area preservation consider the very simple motion of a constant velocity rotor. The two first order equations become

$$d\omega/dt = 0$$

and

$$d\theta/dt = \omega_0.$$

The corresponding phase trajectories are just horizontal lines with differing angular velocities, ω_{0i}, as shown in Figure 2.5. The linear dependence of θ on ω_{0i} ensures that an initial rectangle of points transforms to a parallelogram with a constant base and height, thereby maintaining the original area.

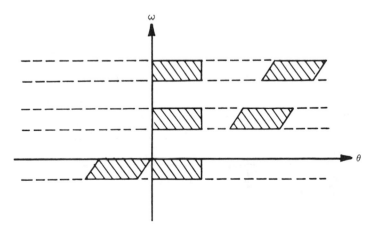

Fig. 2.5 Phase space diagram of a constant velocity rotor.

Example 2.2 also raises the question of boundaries on the phase plane coordinates. In contrast to the linearized pendulum whose finite motion allowed both θ and ω to be bounded conveniently in phase space, the angular coordinate θ for the rotor can increase (positively or negatively) without bound. Yet physically θ is periodic. Therefore the phase diagram is also made periodic by imposing *periodic boundary conditions* on θ as illustrated in Figure 2.6. The θ axis can be limited to $[-\pi,\pi]$, and the two edges of this domain are regarded as identical. As the rotor goes around in the positive θ direction, its phase representation disappears off the right edge of the phase diagram and immediately reappears on the left side. Similar periodic boundary conditions can also be usefully applied to the forced pendulum whose motion passes through the vertical direction.

The property of area preservation, or volume preservation in a higher dimensional space, is a general feature of conservative systems. This property leads to a classification of dynamical systems into two categories – *conservative* or *dissipative* – depending upon whether the phase volumes stay constant or contract, respectively. For example, the linearized undamped pendulum conserves energy, and its trajectories preserve phase area. On the other hand, the trajectories of the linearized damped pendulum,

$$\mathrm{d}^2\theta/\mathrm{d}t^2 + \mathrm{d}\theta/\mathrm{d}t + \theta = 0,$$

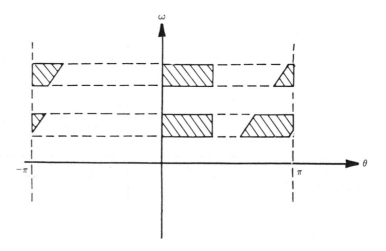

Fig. 2.6 Phase space diagram of the rotor with periodic boundary conditions. Phase points moving to the right disappear at $\theta = \pi$ and then reappear at $\theta = -\pi$.

decay to a single point: $\omega=0=\theta$. This area contraction is illustrated in Figure 2.7. Such a point is called an *attractor*, because a finite set of initial coordinates (θ,ω) converge to it. Obviously phase area is *not* preserved and the system is said to be dissipative.

Using these phase space characteristics we can develop a method for determining from the equations of motion whether a system is conservative or dissipative. The development of the method is easiest to understand in three space dimensions; thus we assume phase coordinates x_1,x_2,x_3, as in Figure 2.8. The equations of motion of the system can be written in terms of the phase 'velocity' components,

$$dx_1/dt = F_1(x_1,x_2,x_3),$$
$$dx_2/dt = F_2(x_1,x_2,x_3),$$
$$dx_3/dt = F_3(x_1,x_2,x_3).$$

Now consider a volume region V with surface S, and assume a net flow of points from V. For a small region ΔS, the flow (or flux) from the region is the component of the velocity vector **v** perpendicular to the surface, multiplied by the element of surface area, ΔS. Since the velocity vector $\mathbf{v} = (dx_1/dt, dx_2/dt, dx_3/dt)$, is specified by the above set of differential equations, then the flux out of the small region is $\mathbf{F} \cdot \mathbf{n}\Delta S$. The net flux out of the entire surface is

$$\text{Flux} = \int_S (\mathbf{F} \cdot \mathbf{n})dS,$$

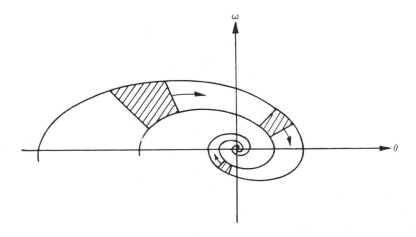

Fig. 2.7 Phase space diagram of the dissipative linear pendulum. Phase area is not preserved.

where **n** is the unit normal vector to the surface, S. Since the flux is the 'velocity' of the phase points from the volume V, then in time δt the indicated volume will change by an amount equal to

$$\delta t \times \text{Flux} = \delta t \int_S (\mathbf{F} \cdot \mathbf{n}) dS.$$

Therefore phase area is preserved or not preserved, depending on the flux integral being zero or negative, respectively. But even this integral will generally be difficult to evaluate, and therefore we simplify the calculation further by using the *divergence theorem* from vector calculus:

$$\int_S (\mathbf{F} \cdot \mathbf{n}) dS = \int_V (\nabla \cdot \mathbf{F}) dV.$$

(See, for example, Thomas and Finney (1979).)

This theorem converts an integral of a vector field such as **F**, on a closed surface S, into an integration of the divergence of **F** over the enclosed volume. Furthermore, the preservation of phase volume should be independent of the particular volume chosen, and it is

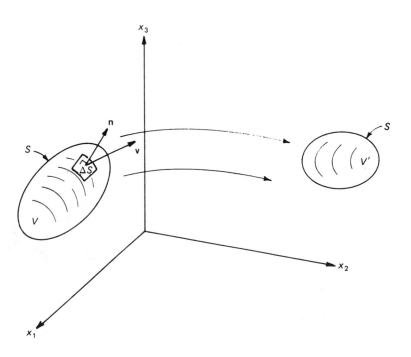

Fig. 2.8 Evolution of a volume in three-dimensional phase space.

therefore sufficient to examine $\nabla \cdot \mathbf{F}$ alone. If this quantity is zero, the system is termed *conservative*, whereas if the divergence of phase velocity is negative the system is *dissipative*. The kinematic properties of the flux in phase space for a conservative system are analogous to the flow of an incompressible fluid in hydrodynamics. (The term *Hamiltonian* is sometimes used in connection with phase volume preserving systems. Many dynamical systems obey Hamilton's equations of motion and such systems are called *Hamiltonian* systems. These systems preserve volume in phase space, according to Liouville's theorem, and therefore Hamiltonian systems are a subset of the set of conservative systems. See Helleman (1983).)

Example 2.3. Let us write both of our example pendula in the phase velocity form and determine their phase space preservation characteristics by this method.

(i) $d^2\theta/dt^2 + \theta = 0$ (undamped) becomes

$d\theta/dt = \omega$ and $d\omega/dt = -\theta$.

Therefore $\mathbf{F} = (\omega, -\theta)$ and $\nabla \cdot \mathbf{F} = \partial\omega/\partial\theta + \partial(-\theta)/\partial\omega = 0$, indicating that phase area is preserved.

(ii) $d^2\theta/dt^2 + d\theta/dt + \theta = 0$ (damped) becomes

$d\theta/dt = \omega$ and $d\omega/dt = -\omega - \theta$.

Therefore $\mathbf{F} = (\omega, -\omega - \theta)$ and $\nabla \cdot \mathbf{F} = \partial\omega/\partial\theta + \partial(-\omega - \theta)/\partial\omega = -1$, indicating that phase area diminishes in time and the system is, as expected, dissipative.

These two examples show how easily the divergence criterion may be applied. For the driven pendulum the equation:

$$d^2\theta/dt^2 + (d\theta/dt)/q + \sin\theta = g\cos(\omega_D t)$$

is converted to a set of first order equations:

$$d\omega/dt = -\omega/q - \sin\theta + g\cos\phi,$$
$$d\theta/dt = \omega,$$
$$d\phi/dt = \omega_D.$$

Then the right sides form the components of the three-dimensional vector **F**. It is left as an exercise for the reader to show that $\nabla \cdot \mathbf{F} = -1/q$ and that, therefore, the system is dissipative.

We have looked at phase diagrams for the damped and undamped linearized pendula. Let us now introduce the full nonlinear restoring torque, $\sin\theta$. Figure 2.9 shows the phase plane for the undamped pendulum:

$$d^2\theta/dt^2 + \sin\theta = 0.$$

For small values of $d\theta/dt$ and θ, the diagram appears similar to that of the pendulum in the linear approximation, but as θ approaches $\pm\pi$ – a pendulum swing that would go all around the circle – the picture changes. At $(\pm\pi, 0)$ the slope develops a discontinuity. The largest of the closed trajectories bounds the region where the motion is oscillatory (or vibrational). On the open trajectory of higher angular velocity, the pendulum goes completely around the circle, and its motion is a rotation modulated by oscillation. The average angular velocity becomes nonzero. (One might compare this to a direct current electrical signal modulated by an alternating current signal.) The corresponding time series of these motions are shown in Figure 2.10.

Consider now the addition of a damping term so the equation becomes

$$d^2\theta/dt^2 + d\theta/dt + \sin\theta = 0.$$

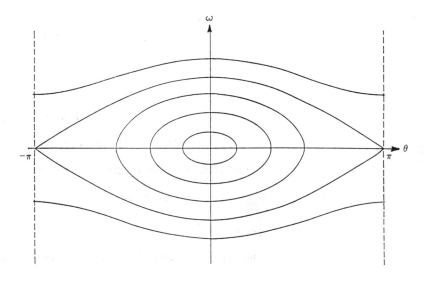

Fig. 2.9 Phase diagram of the nonlinear pendulum. The restoring force term contains $\sin\theta$.

Fig. 2.10 Angular velocity time series. In contrast to the linearized pendulum, the period of the motion for the nonlinear pendulum increases with increasing amplitude. Curves (*a*) and (*b*) show oscillatory motions of differing amplitudes. Curve (*c*) shows the pendulum with sufficient energy to exhibit both rotary and oscillatory motions.

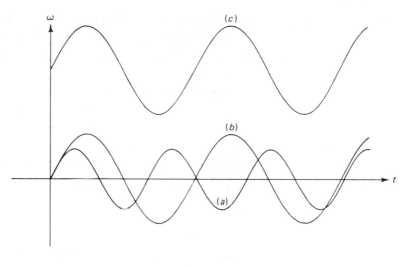

Fig. 2.11 Pair of phase space trajectories for the damped pendulum.

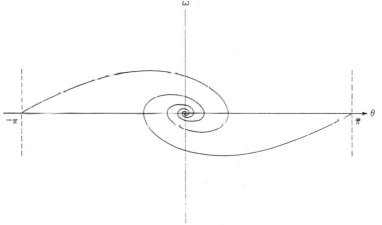

Typical trajectories are shown in Figure 2.11. As indicated previously, the damping term results in an attractor at the origin where $\sin\theta \approx \theta$. Now, however, further attractors are added at $\theta = \pm n\pi$, $\omega = 0$. This can be seen by setting the phase velocity equal to zero and solving for the stationary values of θ and ω; that is,

$$d\theta/dt = \omega = 0$$
$$d\omega/dt = -\omega - \sin\theta = 0.$$

While these attractors are points where the phase velocity goes to zero, questions arise as to the stability of these points. Will the trajectories tend to go back to these critical points if slightly perturbed? Will the stability depend upon the direction of the perturbation? These questions can be answered by looking carefully at the critical points. A useful technique for examining dynamical behavior near critical points involves the assumption that the system will not deviate substantially from linear behavior *near the critical points*. Then each of the nonlinear terms in the differential equations is given a linear approximation near the critical points. This method was developed by Poincaré in 1914 (Hayashi, 1964).

For the case of the damped pendulum,

$$d\theta/dt = \omega$$
$$d\omega/dt = -\omega - \sin\theta,$$

it is easy to see (Problem 6) that near $\theta = \pm n\pi$, where n is even, the linear approximation is

$$d\theta/dt = \omega$$
$$d\omega/dt = -\omega - (\theta - n\pi),$$

and when n is odd the linear approximation becomes

$$d\theta/dt = \omega$$
$$d\omega/dt = -\omega + (\theta - n\pi).$$

In each case θ is transformed to a value centered at the critical point such that $\theta \rightarrow \Delta\theta = \theta - n\pi$ and therefore the linearized phase plane equations are

$$n = \text{even}; \quad d\Delta\theta/dt = \omega \qquad n = \text{odd}; \quad d\Delta\theta/dt = \omega$$
$$d\omega/dt = -\omega - \Delta\theta \qquad\qquad d\omega/dt = -\omega + \Delta\theta.$$

Following the usual method for solving sets of first order linear differential equations, trial solutions of the form $\Delta\theta = Ae^{\lambda t}$ and $\omega = Be^{\lambda t}$ may be substituted into the equations; this yields two pairs of homogeneous algebraic equations. The condition for a nontrivial solution is the vanishing of the determinant of the coefficients of A and B. This condition produces quadratic characteristic equations in λ for each case:

$$\lambda^2 + \lambda + 1 = 0: n = \text{even} \qquad \lambda^2 + \lambda - 1 = 0: n = \text{odd}$$

For the $n = \text{even}$ case, the λ values are complex conjugates with negative real parts. This implies that both $\Delta\theta$ and ω will spiral inward toward the equilibrium point attractor, which is called a *focus*. This

Fig. 2.12 Critical points
in phase space: (*a*) focal
point (*b*) saddle point. In
(*b*) the trajectories going
to the saddle point are
stable whereas the
trajectories coming from
the saddle point are
unstable.

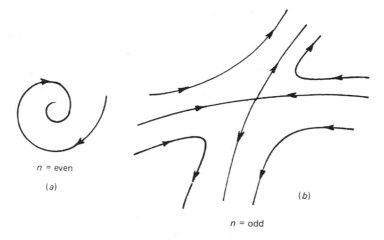

n = even

(*a*)

(*b*)

n = odd

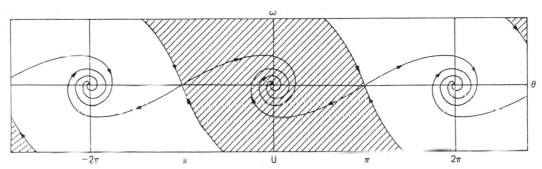

Fig. 2.13 Phase space diagram of the damped pendulum. Alternate shaded and
unshaded regions are basins of attraction. All points within a particular basin are
attracted to the focal point within the basin.

case is shown in Figure 2.12(*a*). On the other hand, the $n=$ odd
condition produces two real values of λ, one positive and one negative.
In this case the stable phase trajectories move toward the critical point
in one direction (negative exponent), but the unstable trajectories
move away from it in another direction (positive exponent). This kind
of critical point is called a *saddle point* and is shown in Figure 2.12(*b*).
The respective directions are obtained by determining the A and B
coefficients appropriate to each λ value. Note how the unstable
trajectory directions correspond to the only possible directions for the
pendulum, located momentarily at the saddle point. The details of the
solution are left as an exercise (Problem 7). Putting all this information together, the phase diagram for the damped pendulum can be
drawn as in Figure 2.13.

This phase diagram suggests yet another property of trajectories in phase space. As drawn in Figure 2.13, the phase space is divided into alternating regions as indicated by the shading. Inside each region all the trajectories will eventually spiral to the enclosed focal point. Each region is the set of all initial conditions (ω,θ) of trajectories that will eventually converge on a specific attractor – in this case a focal point. Such regions are called *basins of attraction*. Furthermore, each of the diagonal curves (Figure 2.13) dividing one basin from another is called a *separatrix*. The arrows on the separatrix (and elsewhere) indicate the flow of the trajectories, toward and away from the saddle points. We will see (Chapter 3) that one characteristic of chaos is the partial dissolution of the separatrix as the basins start to merge.

While most of the discussion so far has focused on the phase *plane*, it is important to realize that the phase space construction need not be confined to two dimensions. A previously introduced equation set,

$$\mathrm{d}x_1/\mathrm{d}t = F_1(x_1,x_2,x_3),$$
$$\mathrm{d}x_2/\mathrm{d}t = F_2(x_1,x_2,x_3),$$
$$\mathrm{d}x_3/\mathrm{d}t = F_3(x_1,x_2,x_3),$$

would define a three-dimensional phase space. We may use this set of equations to illustrate some further properties of phase space. We have already described the divergence method for determining if the above system is conservative. Further, it is evident that the equations are not explicitly time-dependent. The equation set is then called *autonomous* and describes a time-independent flow in phase space, similar to a set of stream lines in a fluid. In fact, the vector **F** is called a *flow*. Autonomous systems also obey the noncrossing property described earlier. However, a projection of a higher dimensional space onto a plane might show apparent crossings which do not represent actual interactions.

The autonomous property is sufficiently useful that it is often desirable to convert a time variable to some other variable in order to make a nonautonomous system into an autonomous system. For example, the variable ϕ is introduced in the driven pendulum equations as $\mathrm{d}\phi/\mathrm{d}t = \omega_\mathrm{D}$ so that the system's dynamical variables become θ, ω, and ϕ. This is convenient since the explicit time dependence enters as a *periodic* term, $g\cos(\omega_\mathrm{D}t)$, and therefore ϕ can be a periodic variable. Then in a three-dimensional phase space, both θ and ϕ can be given periodic boundary conditions such as $\theta \in [-\pi,\pi]$

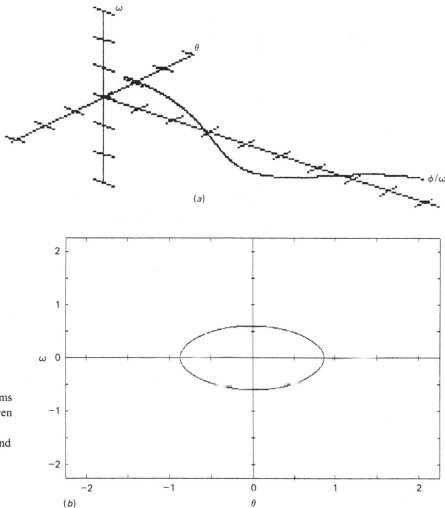

Fig. 2.14 Phase diagrams
for the moderately-driven
pendulum, $g=0.5$:
(a) three-dimensions, and
(b) two-dimensions
($q=2$).

and $\phi \in [0, 2\pi]$. In Figures 2.14(a) and (b), a moderately-driven
pendulum system is illustrated in both three- and two-dimensional
phase spaces. The two-dimensional diagram is a projection of the
three-dimensional diagram. These diagrams show the state of the
pendulum after the initial transient effects have disappeared and after
the system has evolved to a steady state. The resulting closed orbit is
an *attractor* in the same sense as the point is an attractor for the
dissipative, nondriven pendulum. This attractor is obviously one-
dimensional and is called a *limit cycle*.

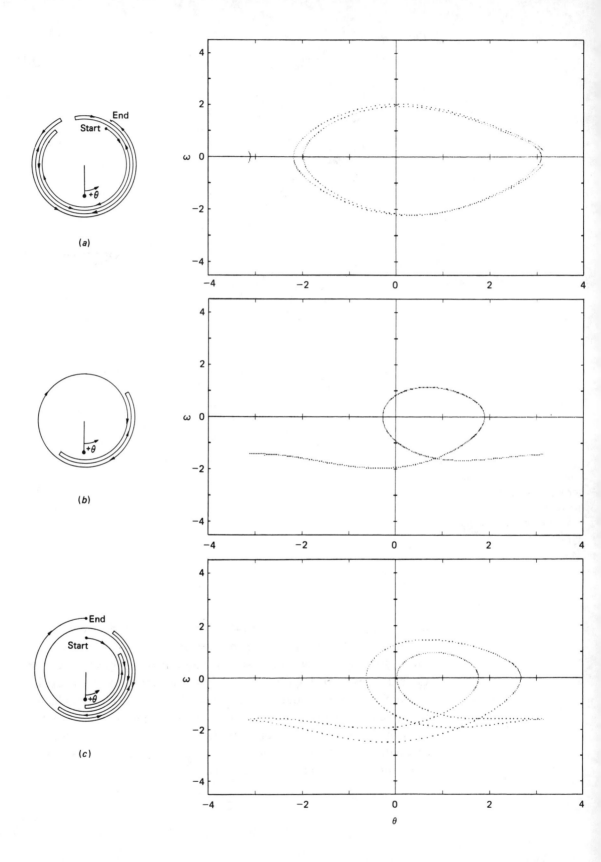

Fig. 2.15 Steady state phase diagram and sketch of the corresponding pendulum motion for different drive amplitudes. All motions are periodic and one complete cycle is shown in each case. The direction of the arrow depends on the initial conditions. (*a*) $g=1.07$; (*b*) $g=1.35$; (*c*) $g=1.45$. ($q=2$).

The motion of the pendulum illustrated in Figure 2.14 is a simple oscillation. As the drive amplitude increases, more complex motions occur, both periodic and chaotic. Some examples of more complex periodic motions are illustrated in Figure 2.15, for various drive amplitudes (g). The path in real space and the corresponding phase plane diagram are shown in each case. The orbits involve a superposition of oscillation and complete rotation.

Poincaré section

A Poincaré section is a device invented by Henri Poincaré as a means of simplifying phase space diagrams of complicated systems. It is constructed by viewing the phase space diagram stroboscopically in such a way that the motion is observed periodically. For the driven pendulum, the strobe period is the period of the forcing.

In order to make this idea more concrete let us refer to the moderately driven pendulum whose attractor was shown in Figure 2.14(*a*). The Poincaré method consists of cutting or sectioning the spiral attractor at regular intervals and looking at these sections along the ϕ axis through the (θ,ω) plane. If this sectioning is done at intervals corresponding to the forcing motion, then the stroboscopic pictures all show one point. The motion always comes back to the same (θ,ω) coordinates as ϕ is increased by 2π. Figure 2.16 illustrates the result.

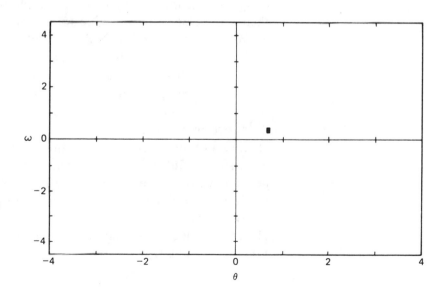

Fig. 2.16 Poincaré section of the linearized pendulum ($g=0.5$, $q=2$, $\phi=0$).

This strobe diagram is called a *Poincaré section* as the graph is 'cut' periodically.

The Poincaré section can provide information about the ratio of the strobe frequency, ω_S, to the natural frequency of the dynamical motion, ω_0. For example, if a motion whose natural frequency was equal to 1 were strobed at a frequency equal to 2, the Poincaré section would have two points. In general, if the natural frequency of the motion, ω_0, is equal to $(p/q)\omega_S$ where p/q is rational, then there are q points, and the order of their appearance is such that, as a given point appears on the circle, the next $[q-(p+1)]$ positions are skipped. All of the q points, however, are eventually filled in. Figure 2.17 illustrates the situation where, for example, when $\omega_0 = \frac{4}{5}\omega_S$, there are five points and $5-(4+1)=0$ positions are skipped as the points go counterclockwise around the circle. These strobe points provide the coordinate values for θ and ω on the Poincaré plot, and the numbering of the points represents their order of appearance.

If the pendulum goes all the way around, then ω has a direct current component as well, and the pattern of dots is not centered on the origin. Since the mixture of rotation and oscillation may lead to a nonzero average displacement $\langle\theta\rangle$, the offset will generally be asymmetric. Furthermore, if the relation between the strobe frequency and the pendulum frequency is irrational (incommensurate), then the strobe points will never quite repeat and the points will gradually fill in a circle on the Poincaré section. Finally, if the system becomes dissipative – in the pendulum case by the addition of a damping term – then the points on the Poincaré section will move toward the appropriate attractor. These ideas are illustrated in Figure 2.18.

For a dynamical system with a periodic forcing term, the Poincaré section provides a simplification of the phase diagram while retaining the essential features of the dynamics. Therefore it is ideally suited to the driven pendulum. In Figure 2.19, some of the periodic motions illustrated by phase diagrams in Figure 2.15 are shown as Poincaré sections. For the periodic motions, the appearance of these sections is quite simple. But in Chapter 3, where the chaotic behavior of the pendulum is described, the simplification of phase space provided by the Poincaré section is shown to be very important for an understanding of the physics. One useful tool for the study of chaos is the observation of the distribution of points on a computer-generated Poincaré section.

Fig. 2.17 Relationship of strobe frequency to dynamical motion frequency. The positioning and order of points on a Poincaré section is shown for different ratios of strobe (ω_s) and motion frequencies (ω_0).
(a) $\omega_0 = \frac{4}{5}\omega_s$; (b) $\omega_0 = \frac{3}{5}\omega_s$; (c) $\omega_0 = \frac{2}{5}\omega_s$.

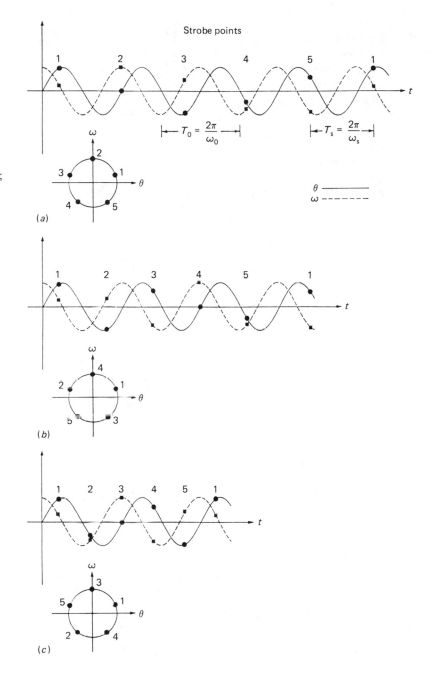

Fig. 2.18 Poincaré
sections of different
motions. (a) Combined
oscillatory and rotational
motion whose frequency
is a rational fraction of
the strobe frequency.
(b) Oscillatory motion
whose frequency is
incommensurate with the
strobe frequency.
(c) Dissipative motion.

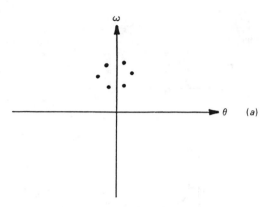

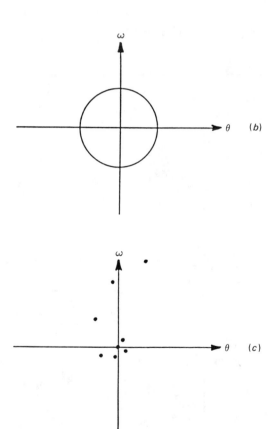

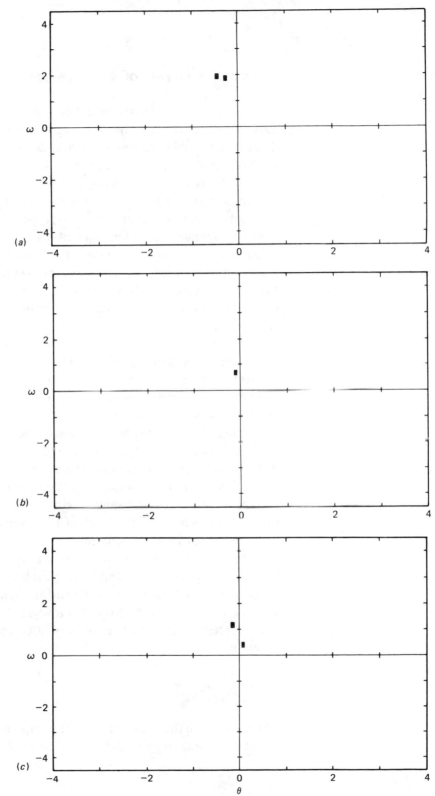

Fig. 2.19 Poincaré sections of motion illustrated in Figure 2.15. The section is taken at $\phi = 0$ $(q = 2)$. (a) $g = 1.07$; (b) $g = 1.35$; (c) $g = 1.45$.

Spectral analysis of time series

The time evolution of a dynamical system is represented by the time variation $f(t)$ or (when sampled at regular intervals) *time series* of its dynamical variables. Any function $f(t)$ may be usefully represented as a superposition of periodic components. The determination of their relative strengths is called *spectral analysis*.

Depending upon the nature of the function, $f(t)$, we may represent it in two different but related ways. If $f(t)$ is periodic, then the spectrum may be expressed as a linear combination of oscillations whose frequencies are *integer multiples* of a basic frequency. This linear combination is called a *Fourier series*. However, it is more likely that $f(t)$ is not periodic, and the spectrum must then be expressed in terms of oscillations with a continuum of frequencies. Such a spectral representation is called the *Fourier transform* of $f(t)$. This representation is especially useful for chaotic dynamics. Because the Fourier transform is in general a complex-valued function, it is often preferable to define a real-valued function which is the modulus squared of the transform. This real function is called the *power spectrum* of $f(t)$. One familiar but crude example of the power spectrum is the LED display of an electronic graphics equalizer. The moving bars on the display indicate the instantaneous electronic power in each of the sections of the audio frequency spectrum.

In this section we review the main features of the Fourier series and then give the Fourier transform method as the limiting case of the Fourier series when the periodicity of $f(t)$ becomes infinitely large, that is, when $f(t)$ ceases to be periodic.

If the function is periodic such that $f(t) = f(t + nT)$ – with n being a positive or negative integer and T being the basic periodicity – then, as noted above, the frequencies of the various spectral components are all integer multiples of the basic frequency, $1/T = \omega_0/2\pi$. The *Fourier series* representation of $f(t)$ may be written compactly in complex notation:

$$f(t) = \sum_{n=-\infty}^{\infty} a_n \mathrm{e}^{in\omega_0 t},$$

where the a_n are the amplitudes of the components of frequency $n\omega_0$. These amplitudes may be determined from the calculation:

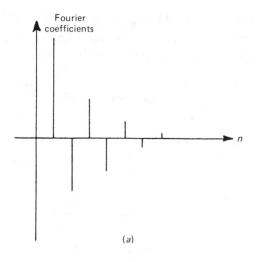

(a)

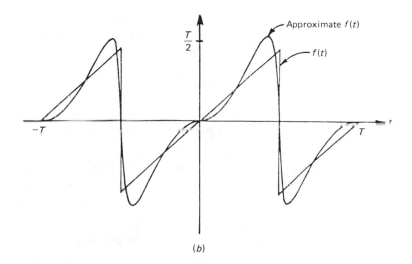

(b)

Fig. 2.20 Fourier spectrum of a sawtooth waveform. (a) Fourier coefficients. (b) Sawtooth, $f(t)$, together with an approximation of $f(t)$ using the two largest components of the Fourier spectrum.

$$a_n = \frac{\omega_0}{2\pi} \int_{-\pi/\omega_0}^{\pi/\omega_0} f(t) e^{-in\omega_0 t} dt.$$

(See, for example, Kaplan (1973).) An example of a periodic time series often found in electronics is the 'sawtooth' function, shown in Figure 2.20.

Example 2.4. The time series of the 'sawtooth' function is $f(t) = t : t \in (-T/2, T/2)$ where $T = 2\pi/\omega_0$. (The pattern is repeated.)

It is left as an exercise to show that

$a_n = 0$ for $n = 0$;
$a_n = 1/in\omega_0$ for $n = $ odd integer; and
$a_n = -1/in\omega_0$ for $n = $ even integer.

Substitution of these results back into the Fourier series expression and manipulation of the ensuing complex expressions leads to the result:

$$f(t) = \frac{2}{\omega_0} [\sin(\omega_0 t) - \tfrac{1}{2}\sin(2\omega_0 t) + \tfrac{1}{3}\sin(3\omega_0 t) - \cdots].$$

Note that the coefficients $1, -\tfrac{1}{2}, \tfrac{1}{3}$, and so forth are not the same as the a_n but are combinations of pairs of a_n. Figure 2.20(*a*) shows a bar chart of the coefficients. The original function $f(t)$, and the resultant of the first two frequency components are shown as an approximation to $f(t)$ in Figure 2.20(*b*).

The *Fourier transform* is an extension of the Fourier series in that the basic periodicity T of $f(t)$ is allowed to become infinitely large. This condition implies that $f(t)$ *need no longer be periodic.* In this circumstance the spacing between the frequency components becomes infinitesimal. The discrete spectrum of frequency components becomes a *continuum* of spectral densities as shown. Therefore, a given component a_n converts to $a(\omega)\delta\omega$ where $\delta\omega$ is a small interval of frequency and $a(\omega)$ is the frequency-dependent amplitude or *Fourier transform*. The practical advantage of the transform is that it can be used to analyze a function about whose properties we are totally ignorant. It often yields surprising and illuminating information.

One may think of the transition from the Fourier series to the Fourier transform in terms of the following set of transformations:

$T \to \infty$
$n\omega_0 \to \omega,$

ω being a continuous variable, and

$a_n \to a(\omega)d\omega.$

Taking the appropriate limits leads to the following conversions:

$$f(t) = \sum_{n=-\infty}^{\infty} a_n e^{in\omega_0 t} \text{ becomes } f(t) = \int_{-\infty}^{\infty} a(\omega) d\omega e^{i\omega t}$$

and

$$a_n = \frac{\omega_0}{2\pi} \int_{-\infty}^{\infty} f(t) e^{-in\omega_0 t} dt \text{ becomes } a(\omega) d\omega = \frac{d\omega}{2\pi} \int_{-\infty}^{\infty} f(t) e^{-i\omega t} dt.$$

The two right hand expressions lead to reciprocal expressions for the Fourier transform,

$$a(\omega) = \frac{1}{2\pi} \int_{-\infty}^{\infty} f(t)^{-i\omega t} dt,$$

and the original function,

$$f(t) = \int_{-\infty}^{\infty} a(\omega) e^{i\omega t} d\omega.$$

As noted earlier, the Fourier transform $a(\omega)$ often turns out to be complex, and it is useful to define a real-valued function, the *power spectrum*, as

$$S(\omega) = |a(\omega)|^2.$$

One might compare this definition with the relation between wave amplitude and wave energy. The power spectrum is the quantity typically calculated in experimental or numerical work.

Let us consider two examples that can be solved analytically.

Example 2.5. Let $f(t)$ be a decaying, oscillating function

$$f(t) = \begin{cases} e^{-\gamma t} e^{i\omega_0 t}, & t \in [0, \infty) \\ 0, & t \in (-\infty, 0] \end{cases}$$

as shown in Figure 2.21(a). The function has a natural frequency ω_0. It could represent a dissipative tuned electrical circuit, for example. Calculation of the integral for $a(\omega)$ leads to

$$a(\omega) = \frac{1}{2\pi[\gamma + i(\omega - \omega_0)]}$$

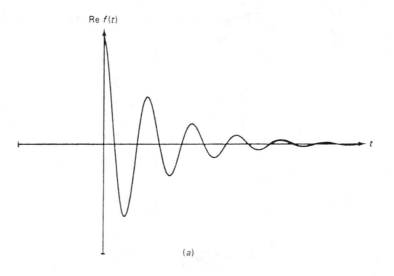

(a)

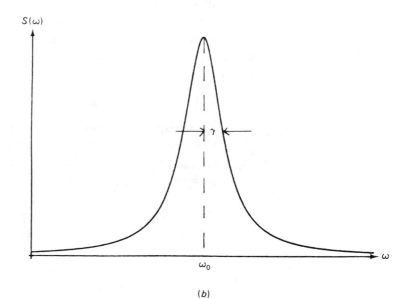

Fig. 2.21 (a) The real
part of the decaying
exponential time series of
Example 2.5. (b) Power
spectrum.

(b)

and then

$$S(\omega) = \frac{1}{4\pi^2[\gamma^2 + (\omega - \omega_0)^2]}.$$

This function is called a *Lorentzian* and is shown in Figure 2.21(b). $S(\omega)$ is symmetric about the dominant 'natural' frequency, ω_0, but because of the damping parameter γ, it has a finite width. If $\gamma \to 0$, then $S(\omega)$ becomes very sharp (approaching a delta function) with all the power then concentrated in the ω_0 component.

There is generally a reciprocal relationship between the width of $f(t)$ and the width of its Fourier transform (or $S(\omega)$). In Example 2.5, the width of $f(t)$ associated with the decaying exponential envelope is $\Delta t = 1/\gamma$, whereas the width of the Lorentzian $a(\omega)$ is $\Delta \omega = \gamma$. The product of the complementary widths is a constant of order unity. As $f(t)$ sharpens in the time domain, $a(\omega)$ broadens in the frequency domain, and vice versa.

Example 2.6. Let $f(t)$ be a wave modulated by a *Gaussian* curve:

$$f(t) = e^{-\frac{t^2}{2\sigma^2}} e^{i\omega_0 t} \qquad t \in (-\infty, +\infty),$$

as illustrated in Figure 2.22(a). Calculation of the integral $a(\omega)$ leads to

$$S(\omega) = e^{-(\omega - \omega_0)^2 \sigma^2},$$

which is Gaussian as shown in Figure 2.22(b). It may be shown (Problem 13) that the width of $f(t)$ is about $\Delta t = \sigma$, and the width of $S(\omega)$ is about $\Delta \omega = 1/(\sqrt{2}\sigma)$, the same reciprocal relationship between the widths holding as in Example 2.5.

Example 2.7. Let $f(t)$ be a linear superposition of two oscillations, with frequencies of $f_1 = 2$ and $f_2 = 3$, with amplitudes of $A_1 = 2$ and $A_2 = 1$, respectively, as illustrated in Figure 2.23(a). The corresponding power spectrum shown in Figure 2.23(b) was obtained numerically, and one easily observes the $4:1$ ratio of the spectral intensities.

Example 2.8. Various types of electrical noise are easily represented by

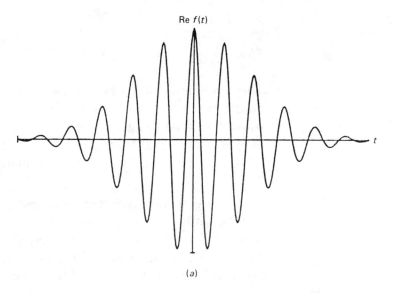

(a)

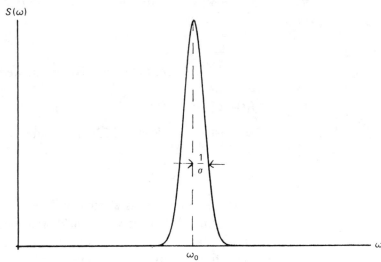

Fig. 2.22 (a) The real part of the Gaussian time series of Example 2.6. (b) Power spectrum.

(b)

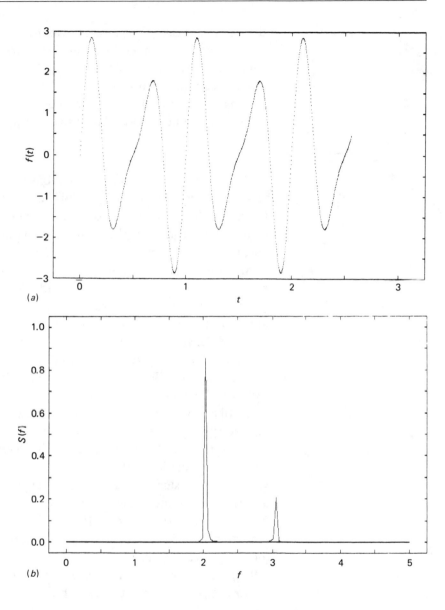

Fig. 2.23 (*a*) Two component times series. (*b*) Power spectrum.

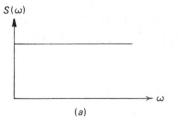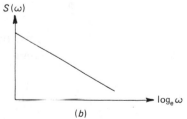

Fig. 2.24 Power spectra of noise time series. (*a*) White noise. (*b*) $1/f$ noise.

their power spectra. For example, Johnson noise which results from thermal agitation in electric circuits is frequency-independent or 'white' (in analogy with white light). On the other hand, '$1/f$ noise, which is common in resistors and solid state devices, has a spectrum varying as f^{-1} (or another power) at low frequencies (Malmstadt, Enke, and Crouch, 1981). These two types are illustrated in Figure 2.24.

Fourier analysis is an interesting subject and the reader who wishes to study it further may find helpful the treatment in Kaplan (1973). Occasionally a brief discussion in the context of quantum physics is provided in the modern physics sections of introductory physics texts. See, for example, Orear (1979).

Analytical calculation of the Fourier transform can become very difficult if the time variation is at all complicated, but numerical methods are straightforward. At first glance, it appears that the appropriate algorithms would involve numerical integration. However, a distinctly different and very efficient approach may be taken when the data are discrete or digitized. The algorithm is called the *fast Fourier transform* (FFT) and was reinvented by J.W. Cooley and J.W. Tukey in 1964. (The method had originally been discovered in 1942 and utilized with hand calculators.) It takes advantage of certain symmetry properties in the trigonometric functions at their points of evaluation, in order to achieve an increase in speed over more conventional methods. If N is the number of data points in the time variation of a signal, then conventional algorithms would require about N^2 computer operations, whereas the FFT requires about $N\log_2 N$ operations. For a 1000 point transform this means a reduction by a factor of about 100, and larger samples lead to even more significant gains. A short introduction is provided in Higgins (1976) and a listing in BASIC is given in Appendix B. (A more comprehensive treatment is available in Press *et al.* (1986), Chapter 12.)

With this background of mathematical tools now available, we may, in Chapter 3, concentrate more fully on the driven, damped pendulum.

Problems

1. For the linearized undamped pendulum show that the solutions, $\omega = A\sin t$ and $\theta = A\cos t$, lead to circular trajectories in the phase plane.

2. Show that one possible solution to the linearized damped pendulum:

   $$d^2\theta/dt^2 + d\theta/dt + \theta = 0$$

   is $\theta = Ae^{-\lambda t}\cos(\omega_0 t)$. Show that $\lambda = \frac{1}{2}$ and $\omega_0 = \sqrt{3/2}$.

3. A particle falls a distance $x(t)$ in a gravitational field, with velocity $v(t)$. The system of equations is

   $$dx/dt = v$$
   $$dv/dt = g$$

   Show that the phase area is conserved. If a friction force, $-kv$, is added to the acceleration equation show that the phase area shrinks.

4. Hénon and Heiles studied stellar orbits in a galaxy using a two-dimensional model with a potential $V(x,y) = \frac{1}{2}(x^2 + y^2) + x^2 y - \frac{1}{3}y^3$. This leads to a system of equations for the four-dimensional phase space of

   $$dp_x/dt = -x - 2xy$$
   $$dp_y/dt = -y - x^2 + y^2$$
   $$dx/dt = p_x$$
   $$dy/dt = p_y$$

   Is this system dissipative?

5. Lorenz developed the following system of equations to describe the interrelations of temperature variation and convective motion

   $$dx/dt = -\sigma x + \sigma y$$
   $$dy/dt = -xz + rx - y$$
   $$dz/dt = xy - bz$$

   where σ, r, b are constants. Prove that the system is dissipative.

6. Show that the system $d^2\theta/dt^2 + d\theta/dt + \sin\theta = 0$ linearizes to $d^2\theta/dt^2 + d\theta/dt + (\theta - n\pi) = 0$ near $\theta = n\pi$ when n is odd, and to

$d^2\theta/dt^2 + d\theta/dt - (\theta - n\pi) = 0$ near $\theta = n\pi$ when n is even. In general one would use a Taylor's series expansion – can this case be treated more intuitively?

7. Find the general solution to $d^2\theta/dt^2 + d\theta/dt + (\theta - n\pi) = 0$ when n is odd.

8. Following the rationale of Figure 2.17 develop the Poincaré plot for

   (i) $\omega = \frac{2}{3}\omega_s$ and (ii) $\omega = \frac{1}{3}\omega_s$.

9. (a) Find the power spectrum of the 'square' wave packet:

   $$f(t) = \begin{cases} a & t \in [0, \pi/2] \\ 0 & \text{for all other values of } t. \end{cases}$$

   (b) Show that the average power is $\overline{f(t)^2} = a^2$,

10. Prove the formula given in the text for the Fourier series amplitudes, a_n.

11. Develop the results for a_n in Example 2.4.

12. Do the calculation of $S(\omega)$ for the decaying exponential wave to obtain the Lorentzian curve.

13. Calculate the width of the Gaussian power spectrum.

The following problems require the use of a computer. Listings provided in Appendix B may be helpful, although they were primarily developed for the driven, damped pendulum, and therefore have to be modified for these exercises.

14. Write a program which will display trajectories in phase space for the undamped, linearized pendulum. The program should require a set of initial (θ, ω) coordinates as input. Remember to keep θ reasonably small so that $\sin\theta \approx \theta$.

15. Write a program which will display trajectories in phase space for the damped, linearized pendulum. Use the equation

   $$d^2\theta/dt^2 + \gamma d\theta/dt + \omega_0^2\theta = 0.$$

16. Modify the program described in Problem 15 so that the term in θ becomes $\omega_0^2\sin\theta$ and try various inputs. In this case you should modify the display of the θ coordinate so that its boundary conditions become periodic, as outlined in the chapter.

17. Modify the program described in Problem 14 so that the input

will be a set of initial coordinates which will form an initial area. The program should then demonstrate the motion of the given area in phase space.

18. Modify the program described in Problem 15 to follow the evolution of a block of initial coordinates. The development of this phase space should illustrate a dissipative system.

19. Develop a program to illustrate Poincaré sections similar to those shown in Figure 2.17. Use the linearized version of the pendulum.

20. In Appendix B the program called EXPFFT computes the power spectrum of a linear combination of periodic components. Use this program to look at $f(t) = \sin(2\pi f_0 t)$. Use a Nyquist frequency of 1 and sample 32 points for $f_0 = \frac{1}{8}$. Try different numbers of points and different Nyquist frequencies. (The Nyquist frequency is the maximum frequency shown by the spectrum.) Try to determine the relationships which involve the number of points, the Nyquist frequency, and the resolution in frequency of the power spectrum.

21. Modify the above program to display the power spectrum of the function c^{-t} on $[0, \infty]$. Experiment with a variety of conditions (Nyquist frequency, number of points, etc.) in your program.

CHAPTER THREE

Visualization of the pendulum's dynamics

Using the tools described in Chapter 2, we are now in a position to discuss the main features of the motion of the driven pendulum. The equations of motion may be written as:

$$\mathrm{d}\omega/\mathrm{d}t = -\omega/q - \sin\theta + g\cos\phi$$
$$\mathrm{d}\theta/\mathrm{d}t = \omega$$
$$\mathrm{d}\phi/\mathrm{d}t = \omega_{\mathrm{D}}$$

Since the system has three variables, its trajectory resides in a phase space of three dimensions, the minimum for chaotic behavior. In this chapter, we present and discuss a variety of computer simulations in order to characterize the dynamics of the pendulum. To allow compact illustration, values of θ and ϕ outside the range $(0,2\pi)$ are plotted at the equivalent point within that range.

The differential equations contain three adjustable parameters: the driving force amplitude g, the damping factor q, and the angular drive frequency ω_{D}. One could define a three-dimensional parameter space in which each point represents a particular choice of the parameters $(g,q,\omega_{\mathrm{D}})$. However a full exploration of the behavior as a function of all three parameters would be a forbidding task. Instead, we fix ω_{D}, choose a few values of q, and let g vary sufficiently to obtain a wide variety of dynamical behavior. As an aid to making appropriate parameter choices, we note that a constant torque of unity is just sufficient to keep the pendulum stationary at $\theta = \pi/2$. Therefore forcing amplitudes in the region $g \approx 1$ are used. Furthermore, the undamped pendulum of small amplitude has a 'natural' angular frequency equal to 1 in our units. Interesting dynamics occur when the forcing-term amplitude is of order unity and the drive frequency ω_{D}

is near (but not equal to) 1. Part of the rich variety of dynamical behavior comes from the interplay between the 'natural' frequency and the drive frequency. One set of parameters containing intervals of chaotic dynamics consists of $q = 2$, $\omega_D = 2/3$, and $0.5 \leq g \leq 1.5$ (Gwinn and Westervelt, 1985).

It is important that the reader develop an understanding of the physical content of the equations and diagrams. One way to accomplish this is to use a computer animation of the pendulum motion. The program in Appendix B entitled MOTION is one version of such a simulation. By running this program or a similar one, the reader can observe the pendulum's behavior for a variety of conditions (especially different values of g). Figure 3.1 shows a 'multiple exposure snapshot' of that animation. We encourage the reader to utilize the simulation frequently while reading this chapter. For some parameter values the motion appears to be periodic, while for others it is chaotic. In some cases the pendulum is nearly periodic for substantial intervals, with intervening irregular intervals; the net effect is that the motion is chaotic.

We are concerned here with the long-term behavior of the pendulum rather than the initial transients, which can be different. Therefore, the animation should be allowed to go through 20 or 30 drive cycles before one considers the motion to have converged to its long-term or statistically stationary state. We will sometimes refer to the long-term behavior as the 'steady state', though the motion need not be time-independent.

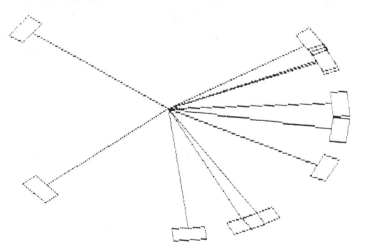

Fig. 3.1 A multiple exposure of the pendulum animation.

Sensitivity to initial conditions

The fundamental characteristic of a chaotic physical system is its sensitivity to the initial state. Sensitivity means that if two identical mechanical systems are started at initial conditions $\mathbf{x}$ and $\mathbf{x} + \varepsilon$ respectively, where ε is a very small quantity, their dynamical states will diverge from each other very quickly in phase space, their separation increasing exponentially on the average. This phenomenon is illustrated for the pendulum in Figure 3.2(*a*). Phase trajectories of a chaotic pendulum, originating at two neighboring points, diverge markedly in less than one forcing period.

Sensitivity may also be illustrated by observing the phase space evolution of a block of pendulum states. (Figures 2.3 and 2.7 display unforced pendula in the undamped and damped cases, respectively.) Figure 3.2(*b*) shows the evolution of a block of initial phase points for the chaotic pendulum. After one half of a forcing period, the initial rectangular block has become long, thin, and curved. Because the system is dissipative, the area of the block shrinks with increasing time. Yet the set of phase points stretches along certain directions and contracts along other directions. The directions of divergence and shrinkage are different at different points in phase space. The net effect is that two closely spaced points are later found quite far apart.

The exponential divergence of adjacent phase points has a further consequence for the chaotic attractor. In order that the trajectories of two adjacent phase points remain bounded without intersecting, they must fold back on themselves, producing a three-dimensional chaotic attractor with many layers (actually an infinite number). A quantitative discussion of exponential divergence and the resulting geometrical complexity of the attractor is given in Chapters 4 and 5.

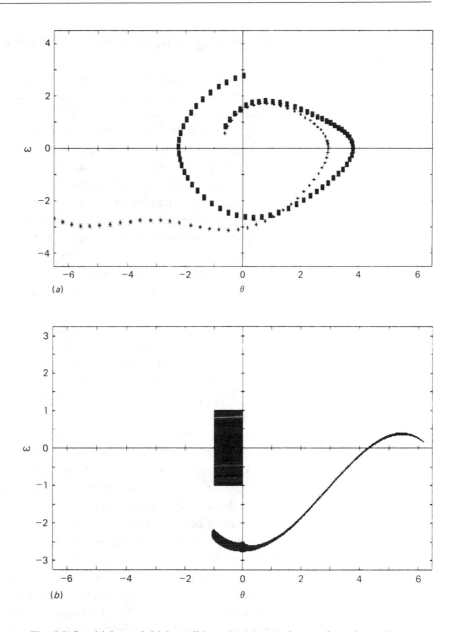

Fig. 3.2 Sensitivity to initial conditions. In (*a*) two phase trajectories, with neighboring initial points near the origin, evolve during one drive cycle ($g = 1.5$, $q = 4$). In (*b*) the phase points of trajectories from a block of initial points, $-0.5 < \theta < 0$ and $-0.5 < \omega < 0.5$, are shown after one half of a drive cycle. (The diagram may be regarded as a projection of the three-dimensional phase trajectories onto the (θ, ω) plane.)

Phase diagrams and Poincaré sections

We now use the geometrical tools of Chapter 2 to characterize the driven pendulum at a variety of driving force amplitudes g. The other parameters are held fixed at $\omega_D = 2/3$ and $q = 2$, though the effect of changing them is also interesting. (These parameters are pure numbers since the pendulum equation was made dimensionless.)

We begin by examining the trajectories in the three-dimensional space $(\theta, \omega, \varphi)$, as shown in Figure 3.3. The first case is periodic, since the trajectory retraces its path exactly. The situations in Figures 3.3 (b) and (c) are clearly more complicated, but it is difficult to tell exactly what has happened from this diagram. Finally, Figure 3.3 (d) is a chaotic state, and the diagram is so complex as to be nearly useless as a way of characterizing the dynamics.

Clearly, a better method of displaying the dynamics is needed. Two-dimensional phase projections and Poincaré sections turn out to be helpful, and these are shown in Figure 3.4. The value of g for each pair of diagrams is given in the caption. The upper parts of Figure 3.4 show projections of the trajectories onto the (θ, ω) phase plane. In this space, periodic motion appears as a closed orbit. Of course the projected orbits can appear to cross, and this occurs in the more complicated cases shown.

The lower parts of Figure 3.4 are the Poincaré sections, which are simply slices across the φ axis of the three-dimensional attractor. Periodic orbits $((a), (b), (d), (e), (f))$ appear as a finite number of dots (enlarged for clarity), while chaotic orbits $((c), (g))$ form complicated sets containing an infinite number of points. We shall return to an examination of their structure shortly.

The shape of the Poincaré sections varies with the phase at which they are taken. Sections for different values of φ are shown in Figure 3.5, and the aggregate of these shapes is similar to the full attractor of Figure 3.3 (d). As φ is increased, the attractors become stretched and folded repetitively, much like the kneading of dough. Evidence of this stretching and folding process may be seen in the fact that the sections contain a number of layers.

Actually, the structure of the attractors is much more complicated than is apparent from the sequence of Poincaré sections in Figure 3.5. This may be illustrated by looking at a small part of one of the sections, greatly magnified, as shown in Figure 3.6. The three parts of

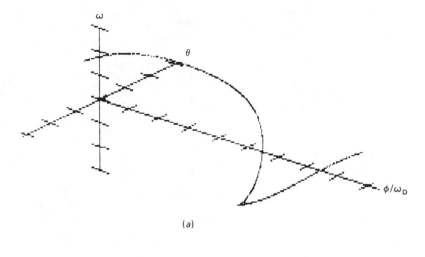

(a)

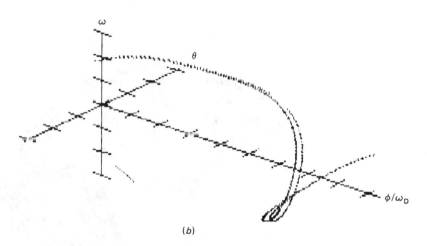

(b)

Fig. 3.3 Three-dimensional phase portraits for several values of driving force strength and $q=2$. (a) $g=0.9$; (b) $g=1.07$; (c) $g=1.47$; (d) $g=1.5$. This last case is chaotic.

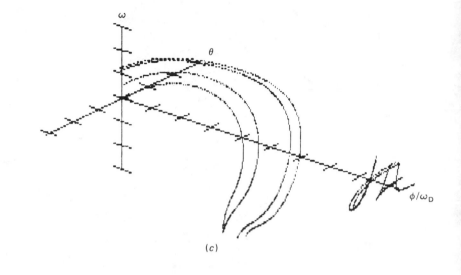

(c)

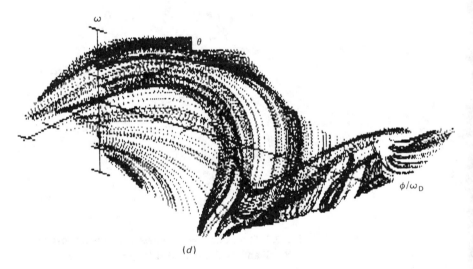

(d)

Fig. 3.3 (*cont.*)

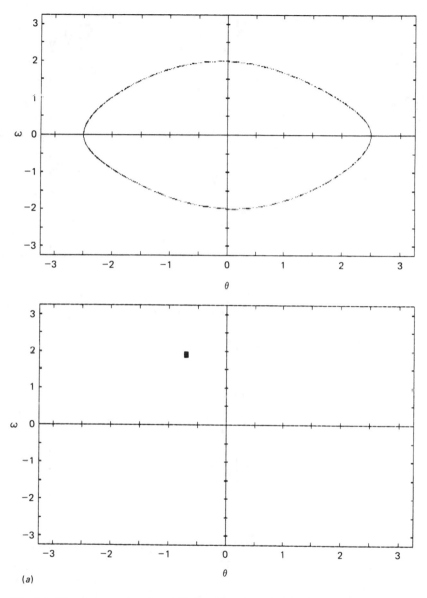

(a)

Fig. 3.4 Phase plane (above) and Poincaré sections (below) for several values of driving force amplitudes and $q=2$. In some cases the dots of the Poincaré sections have been enlarged for clarity. (a) $g=0.9$. (b) $g=1.07$, and a period doubling is apparent; (c) $g=1.15$, and the system is chaotic; (d) $g=1.35$, and the system is periodic again; (e) $g=1.45$, and another period doubling has occurred; (f) $g=1.47$, and a second period doubling is apparent; (g) $g=1.50$, another chaotic state.

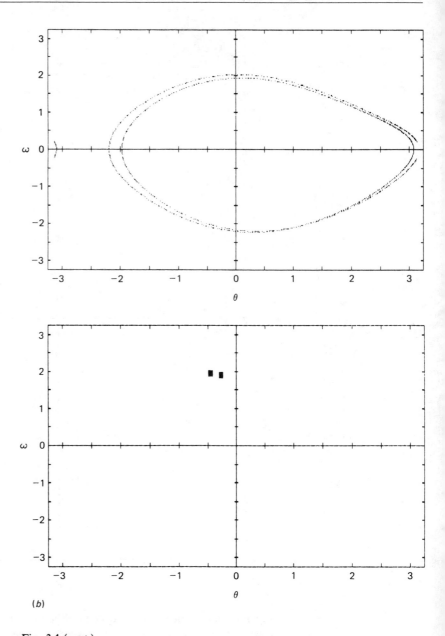

(b)

Fig. 3.4 (*cont.*)

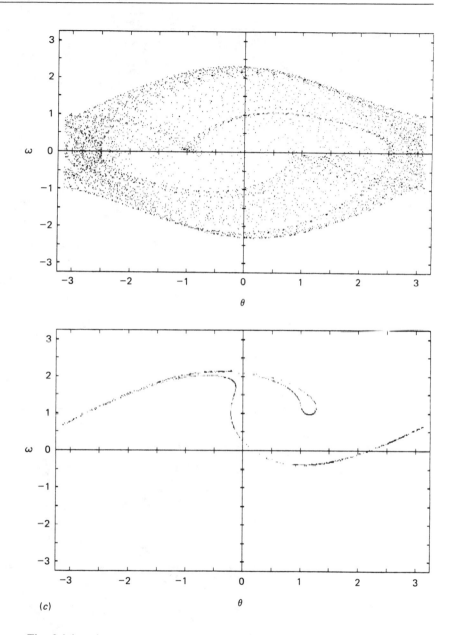

(c)

Fig. 3.4 (*cont.*)

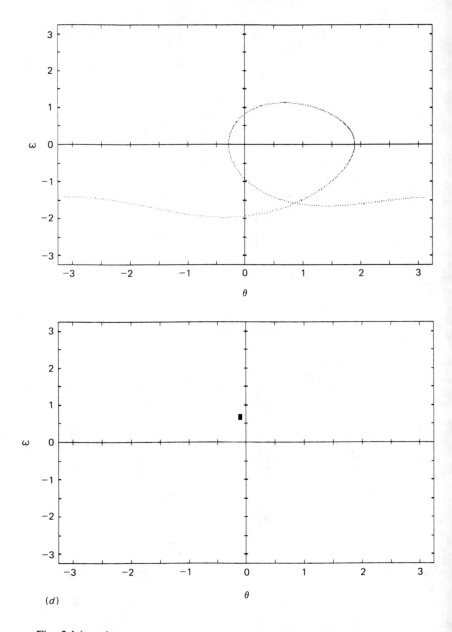

(d)

Fig. 3.4 (*cont.*)

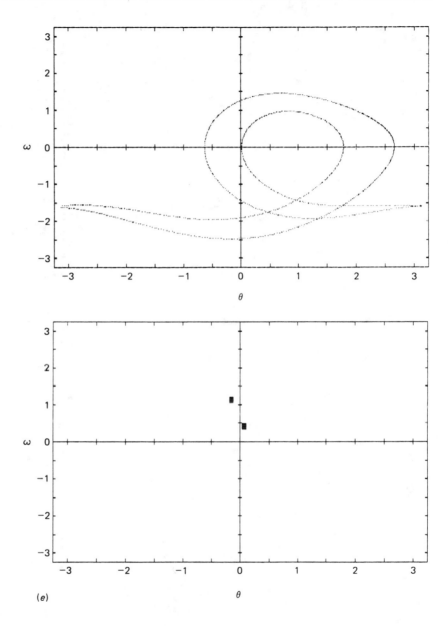

(e)

Fig. 3.4 (*cont.*)

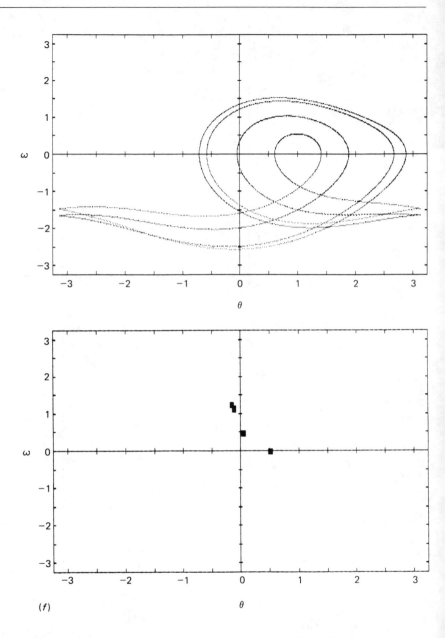

(f)

Fig. 3.4 (*cont.*)

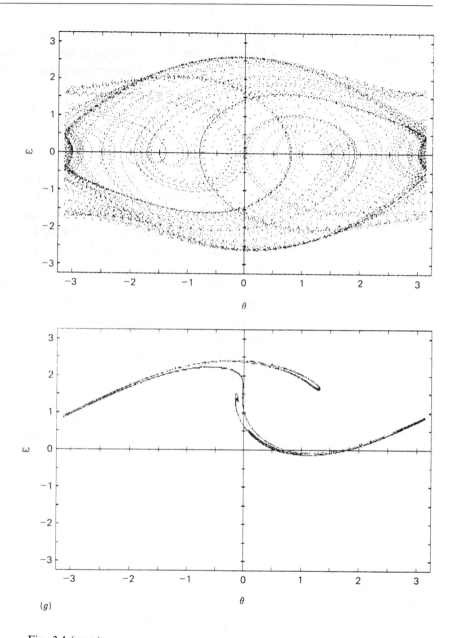

Fig. 3.4 (*cont.*)

this diagram show the attractor at different scales of magnification. (The different scales are obtained simply by changing the window for the graph. Each graph is abstracted from a set of about 10 000 points.) The stretching and folding processes lead to a cascade of scales: the attractor consists of an infinite number of layers. The fine structure, when magnified, resembles the gross structure. This property is called *self-similarity*.

These simulations of the chaotic attractor and its Poincaré sections reveal a hierarchical structure that is uncharacteristic of ordinary compact geometrical objects. In Chapter 5 the chaotic attractor and corresponding Poincaré sections are discussed as *fractals* – mathematical sets of noninteger dimension. While the periodic attractor is one-dimensional (Figures 3.3 (*a*), (*b*) and (*c*) and its Poincaré section is zero-dimensional (a few points), chaotic attractors are more complex, and their dimension is a fraction greater than two (see Chapter 5). Attractors having noninteger dimension are called *strange attractors*.

Finally, in Figure 3.7, we show several chaotic Poincaré sections corresponding to different values of the damping coefficient q. The layers of the attractor are more widely spaced as the damping decreases (or q increases).

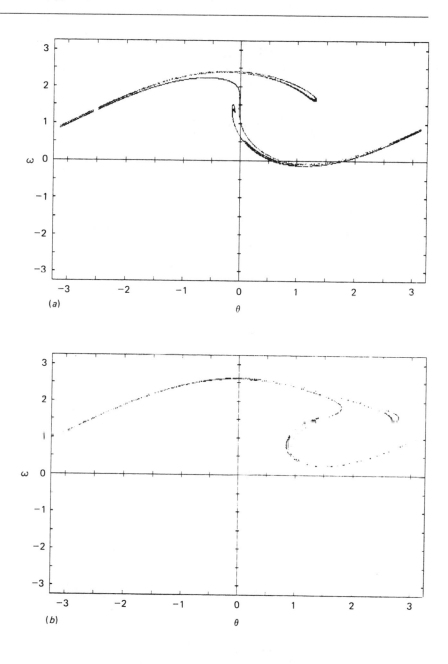

Fig. 3.5 Poincaré sections taken at incremented values of ϕ, the phase of the forcing term. $\Delta\phi = 2\pi/10$. At $\phi = \pi$ the section is anti-symmetric to the $\phi = 0$ case. $g = 1.5$, $q = 2$. (a) $\phi = 0.0$, (b) $\phi = 0.628319$, (c) $\phi = 1.25664$, (d) $\phi = 1.88496$, (e) $\phi = 2.51327$, (f) $\phi = 3.14159$.

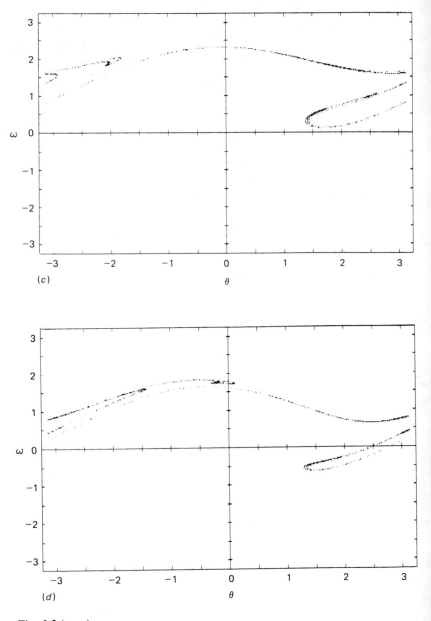

Fig. 3.5 (*cont.*)

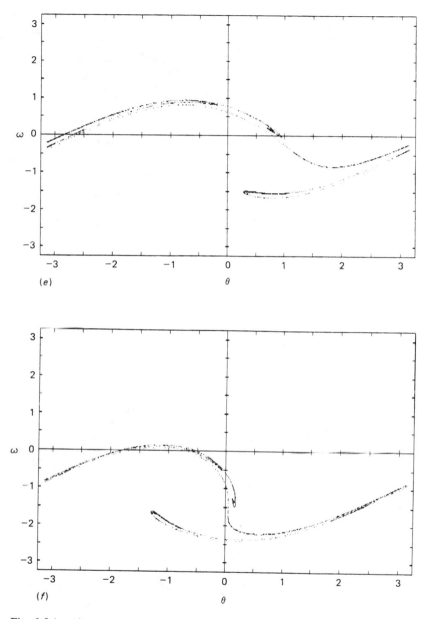

Fig. 3.5 (*cont.*)

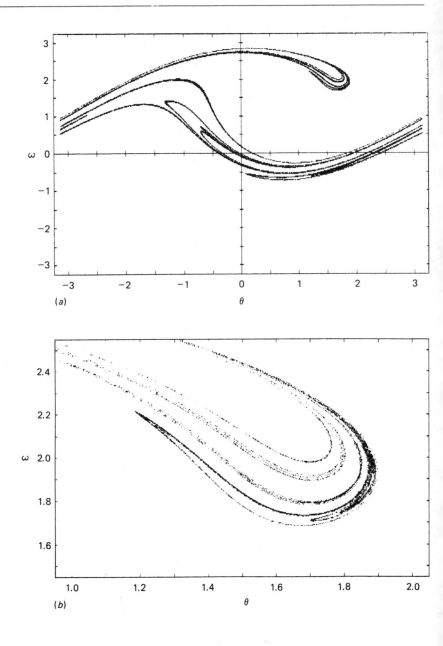

Fig. 3.6 (*a, b, c*) Attractor in the Poincaré section for $q = 4$ and $g = 1.5$ viewed at different magnifications, thus revealing the self-similar structure caused by the folding and stretching of phase volume.

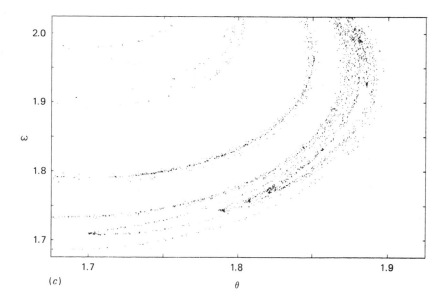

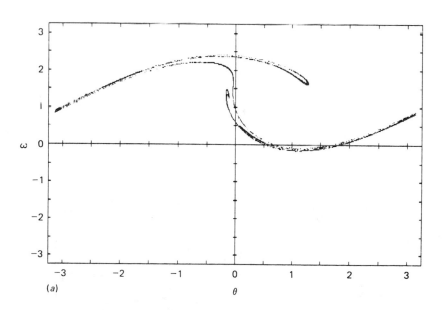

Fig. 3.7 Attractors in the Poincaré section for chaotic states of pendula with different amounts of damping ($g=1.5$): (a) $q=2$, (b) $q=2.8$, (c) $q=4$.

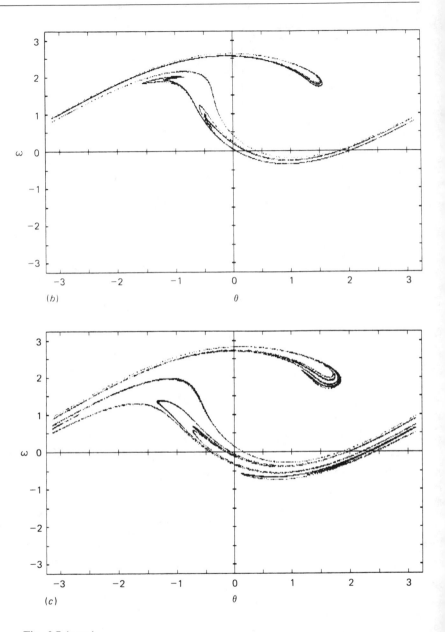

Fig. 3.7 (*cont.*)

Time series and power spectra

In Chapter 2 the power spectrum was introduced as a representation of the relative abundance of different frequencies in a given time series. Figure 3.8 shows a time series and power spectrum for the angular velocity ω at a drive amplitude of $g = 0.95$. The time series in Figure 3.8(a) shows a periodic oscillation. The corresponding power spectrum (Figure 3.8 (b)) exhibits a strong peak at the drive frequency, $1/(3\pi)$ together with some higher frequency harmonics. The harmonics are not unexpected since the phase space pattern is asymmetric. Logarithmic plots are used to highlight components with low power levels – an important feature of chaotic spectra.

In Figure 3.9 a corresponding set of diagrams is given for a chaotic state at $g = 1.5$. The time series is obviously irregular. The power spectrum is broadband, and contains substantial power at low frequencies. A sharp component at $\omega_D/2\pi$ is also present. Though a broad spectrum does not guarantee sensitivity to initial conditions, it is, in practice, a reliable indicator of chaos.

This book is primarily concerned with dynamical systems defined by sets of differential equations. However, it is worth noting that power spectra are also very useful for the analysis of experimental data. Measurements typically include time series of some dynamical variable, and the corresponding power spectrum can be readily analyzed to determine the state of the system. (See, for example, Gollub and Benson (1980) and Iansiti *et al.* (1985).)

Another useful technique for distinguishing chaotic and nonchaotic motions is the calculation of *Lyapunov exponents*, which are quantitative measures of the evolution of neighboring phase trajectories. As with Fourier analysis, the method is applicable to both numerical and experimental data; we describe it in Chapter 5.

Basins of attraction

Figure 2.13 showed the phase portrait of the damped *unforced* pendulum. Each of the point attractors at $\theta = 2n\pi$ (n = integer) is encompassed by a region called a *basin of attraction*. All the points (θ, ω) in the basin converge on the enclosed point attractor. The boundary between two basins of attraction is called the *separatrix*.

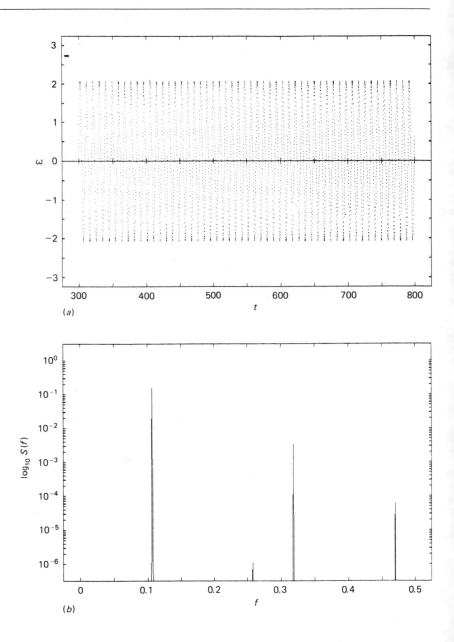

Fig. 3.8 (*a*) Time series and (*b*) power spectrum of angular velocity, ω, for periodic motion at $g = 0.95$.

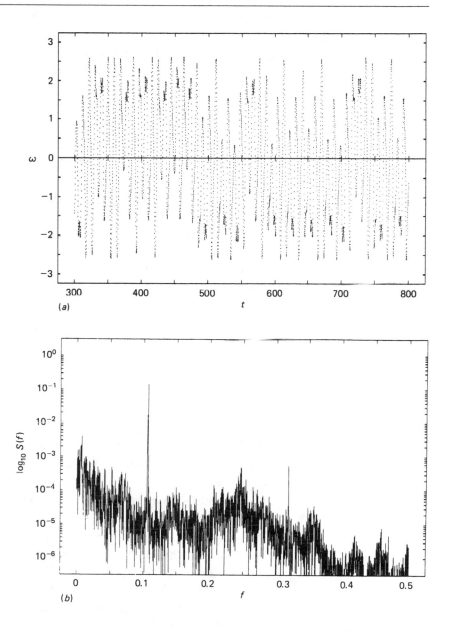

Fig. 3.9 (*a*) Time series and (*b*) power spectrum of angular velocity for chaotic motion at $g = 1.5$. The peak is located at the drive frequency.

For the case cited, the separatrix is a line defined by the stable phase trajectory going to the saddle point as shown in Figure 2.13.

In order to determine basins of attraction for the *forced* pendulum numerically, it is necessary to take advantage of some property that differs from one basin to another. For example (Gwinn and Westervelt, 1985), one can use the fact that, in the region $g > 1.3$, there are two stable rotary modes with average components of angular velocity close to $\pm \omega_D$ for the different basins. The phase portraits of these modes are shown in Figures 3.10(a) and (b).

The basins of attraction are obtained by taking each pair (θ, ω) of initial conditions on a grid, and calculating the trajectory of that pair over many cycles. To eliminate transient effects, the first 30 cycles are discarded; and the velocity is then averaged over the remaining cycles. The two basins of attraction are distinguished by the *sign* of $\langle \omega \rangle$ and, for positive $\langle \omega \rangle$ a circle is put at the corresponding location of the initial condition. Figure 3.11 shows the basins of attraction for $g = 1.3$, a periodic state.

On a large scale the basins of attraction of Figure 3.11 bear some resemblance to those of the undriven pendulum, but the basin boundaries appear fuzzy. In fact, careful studies (Gwinn and Westervelt, 1986) have shown the boundaries to be fractals (see Chapter 5); that is, the basins are interwoven near the boundaries. If the initial phase space coordinates of a trajectory near the boundary are not specified precisely, the basin of attraction for the trajectory is uncertain. This uncertainty is related to the fractal dimension of the boundary.

Further insight may be gained by looking at the basins of attraction with the Poincaré section superposed (Gwinn and Westervelt, 1985). This type of diagram may be generated by insertion of the Poincaré algorithm into the basin of attraction program. For a nonchaotic state one finds that each piece of the Poincaré section is unambiguously inside a single basin of attraction, as in the case $g = 1.47$. (See Figure 3.12(a) where the Poincaré attractor consists of eight points, four from each of the two sets of initial conditions.) For $g = 1.48$, Figure 3.12(b), the attractors spread out, reaching toward the basin boundaries. Finally, in the chaotic state for $g = 1.5$, the basin structure breaks up, and the previously separate attractors corresponding to two different initial conditions join together to form a single attractor consisting of an infinite number of lines, as shown in Figure 3.12(c).

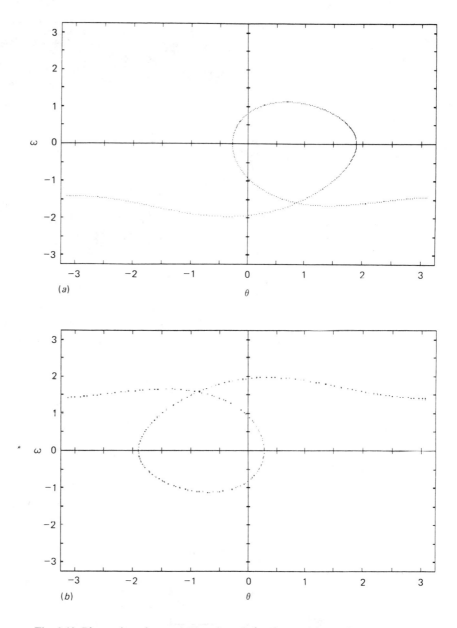

Fig. 3.10 Phase plane for $g = 1.35$ and $q = 2$ showing positive and negative drifting states for different initial conditions: (a) $\theta_0 = 0$, $\omega_0 = 0$; (b) $\theta_0 = 0$, $\omega_0 = 2.3$.

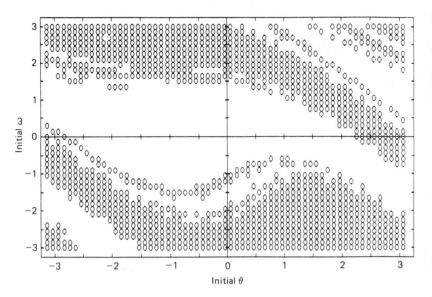

Fig. 3.11 Basin of attraction for $g = 1.3$, a state of periodic motion. The circles indicate positive drift of the angular velocity. The blank regions correspond to negative average angular velocity. The basins are intertwined.

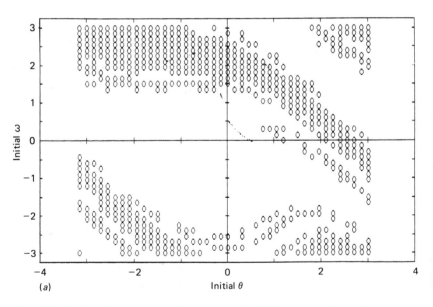

(a)

Fig. 3.12 Basins of attraction with the superposed Poincaré sections. The open circles indicate a positive average $\langle \omega \rangle$. (a) The Poincaré section consists of small clusters of points when $g = 1.47$. (b) The attractors have spread toward the basin

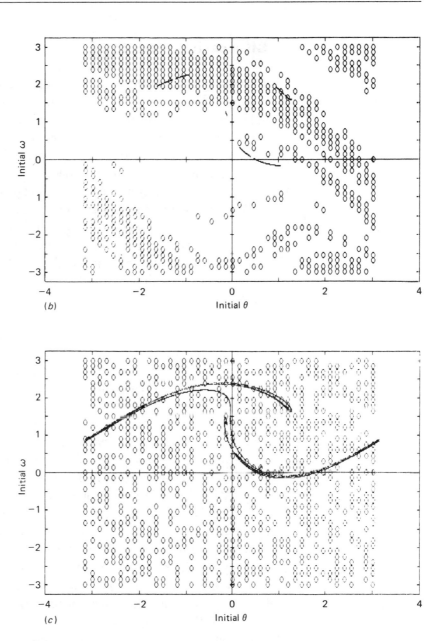

(b)

(c)

boundaries for $g = 1.48$. (c) The separate attractors corresponding to two different initial conditions merge and the basins lose their identities at the onset of chaos. Here, $g = 1.5$. In each case $q = 2$.

Bifurcation diagrams

Phase diagrams, Poincaré sections, time series, and power spectra provide information about the dynamics of the pendulum for specific values of the parameters g, q, and ω_D. The dynamics may also be viewed more globally over a range of parameter values, thereby allowing simultaneous comparison of periodic and chaotic behavior. The bifurcation diagram provides a summary of the essential dynamics and is therefore a useful method of acquiring this overview.

For some values of the parameters, a pendulum will have only one long-term motion, while for other slightly different choices, two or more motions may be possible. If several of them are stable, the actual behavior will depend on initial conditions. In dynamics a change in the number of solutions to a differential equation as a parameter is varied is called a *bifurcation*.

For the pendulum, bifurcations can be easily detected by examining a graph of ω (at a fixed phase in the drive cycle) versus the drive amplitude g. Several examples of these graphs, called *bifurcation diagrams*, are shown in Figure 3.13. The interpretation is relatively straightforward. If the pendulum is lightly driven and the motion is periodic with the same period as the drive frequency, ω_D, then the angular velocity ω has one value at a given time (point of constant phase) during the drive cycle. If the parameter g is increased sufficiently, further components of longer period are added to the motion, and one observes more than one value of ω at the given phase. The system has undergone a bifurcation.

For the diagrams shown in Figure 3.13, ω is taken at the *beginning* of the drive cycle ($\phi = 0$). The system is allowed to come to a steady state by omitting the first 30 drive cycles. The figure shows the next 30 drive cycles. Suppose first that the pendulum is lightly driven (say $g = 0.9$), as in Figure 3.13(*a*). Its motion is an oscillation at the forcing frequency. The phase trajectory is a limit cycle that is symmetric about the origin; the corresponding Poincaré section shows a fixed point. The angular velocity takes only a single value in the bifurcation diagram.

If the driving force is slightly increased to about 1.025, then the phase trajectory loses its symmetry about the origin and has two different shapes (Figure 3.14) depending on the choice of initial

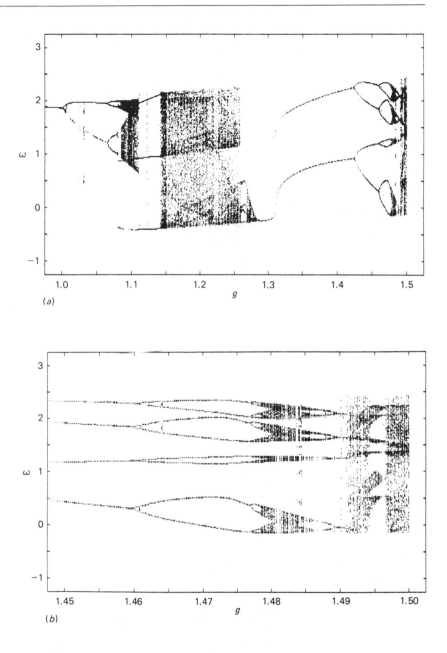

Fig. 3.13 (*a*) Bifurcation diagrams showing the long-term values of the angular velocity ω at the beginning of each drive cycle, plotted against the forcing amplitude *g*. (*b*) An expansion of one region of (*a*). The other parameters are $q = 2$ and $\omega_D = 2/3$.

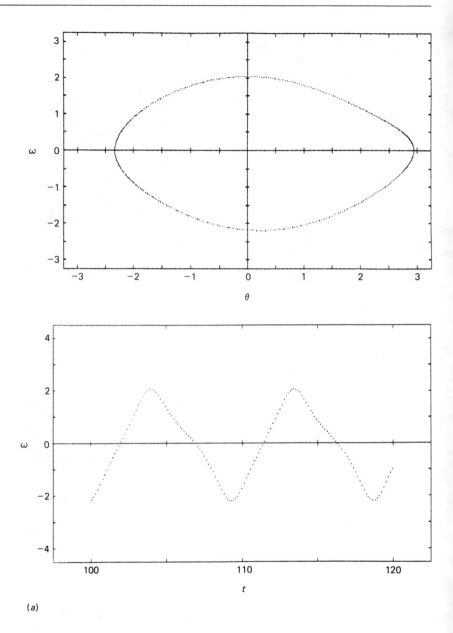

Fig. 3.14 Phase diagrams (above) and velocity time series (below) showing two periodic trajectories (a), (b), which develop from different sets of initial conditions. $g = 1.025$, $q = 2$, $\omega_D = \frac{2}{3}$.

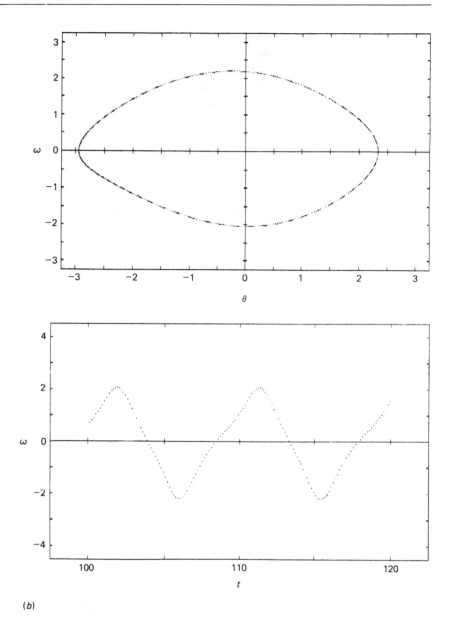

(b)

conditions. This two-valuedness appears in Figure 3.13(a) as a splitting. But note that the pendulum's motion still is oscillatory, with a main frequency ω_D and possibly some higher frequency harmonic content. Each set of initial conditions produces only one value of ω in the bifurcation diagram.

If the driving amplitude is increased to approximately 1.07, the periodicity of the pendulum doubles, and it now has frequency components at ω_D and $\omega_D/2$. Observation of the animation for a given set of initial conditions shows two slightly different oscillatory motions of frequency ω_D whose combination has a frequency of $\omega_D/2$. The motion is sketched in Figure 2.15(a). This effect is called *period doubling*. It causes the system to cycle between two values of ω (at the beginning of the drive cycle) for each set of initial conditions. This change is evident in the bifurcation diagram of Figure 3.13(a). Given the two-valuedness resulting from the two asymmetric attractors, a total of four values of ω may occur at $\phi = 0$.

The bifurcation diagram is very complex. For certain ranges of the parameter g, the angular velocity takes on an infinite number of values, though there are also many holes; these states are chaotic. It is also interesting to see that within the chaotic regions there are small intervals in which the motion abruptly becomes periodic again (for example, $g \sim 1.12$). Beyond the large chaotic region occupying much of the interval $1.08 < g < 1.28$, a wide interval of periodic motion appears again, centered at $g = 1.35$. Study of the animation in this region shows a rotary motion with a small, superposed oscillation. Depending upon the initial conditions, the rotary motion has either a positive or negative average angular velocity. (See Figure 3.3(e).) This two-valuedness is evident in the bifurcation diagram.

Beyond $g = 1.43$, a new *subharmonic cascade* occurs. At $g = 1.45$ (see Figure 3.13(b)) the Poincaré section consists of two points, and at $g = 1.47$ four points (for a given set of initial conditions). In the region around $g = 1.48$ there are four densely occupied bands of ω. The motion is chaotic and ω takes different values in a regular cycling around the bands. A narrow periodic interval occurs for $1.487 < g < 1.493$, followed by chaotic motion for higher g.

The bifurcation sequences observed as a function of g change dramatically if the parameters q and ω_D are changed. One example of a different sequence is given in Figure 3.15, where the damping factor is reduced by a factor of 2 so that $q = 4$. The regions of chaotic behavior are much broader, and there is a prominent window of periodic behavior around $g = 1.25$.

The bifurcation diagram is an important tool for discovering interesting parameter regimes for a dynamical system. While our discussion focused upon variation of the forcing g, bifurcation diagrams utilizing q and ω_D as the independent variable also yield similar displays of varied dynamical behavior.

The numerical computations discussed in this chapter illustrate the complexity and variety of motions of the pendulum. Experiments illustrating some of these phenomena have been performed (Blackburn *et al.*, 1989). Analytic solution of the pendulum equations is apparently not feasible except for special cases. Therefore, in order to understand the development of chaos we consider in Chapter 4 several nonlinear mappings, which are more tractable than differential equations. Despite their simplicity, these maps exhibit many of the phenomena illustrated by the pendulum.

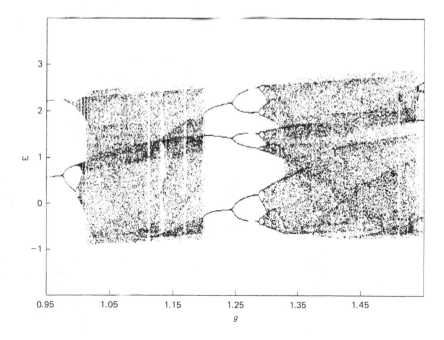

Fig. 3.15 A bifurcation diagram for $q = 4$; this corresponds to lighter damping than in Figure 3.13.

Simulations

1. Use the program MOTION listed in Appendix B or the MOTION option in the menu of the program CHAOS to study the motion of the driven pendulum. Try different values of the parameter set (ω, g, q). Let the motion run for many cycles in order to observe the long term behavior.

2. Use the program PENDULUM listed in Appendix B or the 2D-PHASE DIAGRAM option in the menu of CHAOS to study the two-dimensional projection of the phase space for the pendulum. Try different values of (ω_D, g, q) and different sets of initial conditions (θ_0, ω_0). Discard the first ten cycles to allow the pendulum to reach a steady state.

3. Use the INIT. BLOCK FLO option in the menu of CHAOS (or modify PENDULUM) to study the motion of a block of initial conditions in (θ, ω) space. Note the change in shape and area of the block.

4. Use the program POINCARE listed in Appendix B or the POINCARE SECTION option in the menu of CHAOS to study the Poincaré section of the pendulum. Try different values of (ω, g, q) and phase angle. Discard the first ten cycles to allow the pendulum to reach a steady state.

5. An electrical circuit with resistance, inductance, and nonlinear capacitance may be driven sinusoidally into chaotic states. The differential equation for the circuit is

$$d^2x/dt^2 + A\,dx/dt + x^3 = B\cos t$$

where A and B are adjustable parameters. It has been suggested that the transition to chaos may be observed for parameter values $A = 0.1$ and $9.8 < B < 13.4$ (Moon, 1987, p. 272). Modify the programs PENDULUM and POINCARE (listed in Appendix B) or the source code on the diskette for the libraries LPHASE2D and LPOINCAR in order to develop programs to study the behaviour of this dynamical system. Note that the drive angular frequency is 1. Eliminate the periodic boundary conditions on the position coordinate and put larger boundaries on the axes.

6. Use the listing BIFURCATION in Appendix B or the BIFUR-CATION DIAG option in the menu of CHAOS to generate a

bifurcation diagram for the pendulum. Since the process is equivalent to the computation of many Poincaré sections, the program takes a lot of processing time and you may wish to save the computed data on a separate diskette.

7. Modify the bifurcation program to develop a bifurcation diagram for the equation of Problem 5. Use the range of B values suggested in that problem.

8. Use the FFT listing in Appendix B or the FFT option from the CHAOS menu to generate a power spectrum for the pendulum for $g = 1.5$ and $q = 4$.

9. Modify the Runge–Kutta procedure in the FFT program for the equation of Problem 5 and run the program for a value of B which gives a chaotic behavior. (Use your bifurcation diagram from Problem 7 to find an appropriate value of B.)

10. Use the BASINS listing in Appendix B or the BASINS OF ATTRA option from the CHAOS menu to generate a diagram of the basins of attraction. Try $g = 1.3$ and $q = 2$.

CHAPTER FOUR

Toward an understanding of chaos

The driven pendulum, our primary example in this work, may seem to be one of the simplest physical systems to exhibit chaotic behavior. Mathematically and computationally, however, nonlinear differential equations are difficult to solve. Even more elementary model systems can give insight into the mechanisms leading to chaotic behavior. These are stated in the form of *difference equations*, rather than *differential equations*. A typical difference equation is of the form

$$x_{n+1} = f(\mu, x_n),$$

where x_n refers to the nth value of x, a real number on the unit interval $(0,1)$, and μ is a parameter. One may think of nT as a time, where T is a basic time interval. The parameter μ may vary with the particular model and, in the examples we will discuss, varying μ leads to the onset of chaotic behavior. The function, f, is said to be a *map* of the interval $(0,1)$ onto itself, since it generates x_{n+1} from x_n. The function $f(\mu, x_n)$ may be nonlinear in the argument x_n, just as the differential equations for the pendulum are nonlinear in θ.

Difference equations may be solved quite readily by iteration, and their numerical solution is much less time consuming than is the case for differential equations. A dynamical system whose phase space is three-dimensional (such as the pendulum) may be converted to a mapping through the Poincaré section. These sections are 'snapshots' of the phase plane taken at each drive period. Because the differential equation is deterministic a definite relation exists between the coordinates (θ_n, ω_n) at the end of the nth drive period and the coordinates $(\theta_{n+1}, \omega_{n+1})$ at the end of the $(n+1)$th period. This relation can, in principle, be written as a mapping in two dimensions:

$$\theta_{n+1} = G_1(\theta_n, \omega_n)$$
$$\omega_{n+1} = G_2(\theta_n, \omega_n).$$

Sometimes (but not generally) such mappings may be further simplified to one dimension. While a particular analytic form of the pendulum mapping has not been found, simple maps may be used to illustrate some aspects of the pendulum's behavior.

Because of their relative simplicity, one-dimensional maps provide several advantages over the differential equations. They allow for simple, clear statements of many characteristics of chaotic behavior, such as sensitivity to initial conditions and the evolution of information. Maps also illustrate clearly the mechanisms of bifurcation of solutions, and the folding and stretching required for chaos in a finite phase space. In this chapter several maps and their properties are explored to aid in further understanding chaotic behavior.

The logistic map

This simple map, given by the difference equation

$$x_{n+1} = \mu x_n(1 - x_n), \qquad x_n \in [0,1]$$

takes its name from the corresponding differential equation

$$dx/dt = \mu x(1 - x)$$

originally used by P.F. Verhulst in 1845 to model population development in a limited environment (May, 1976). The logistic map is one-dimensional and nonlinear, and may be visualized as indicated in Figure 4.1. The diagram has three parts: the parabolic curve $y = \mu x(1 - x)$, the diagonal line $x_{n+1} = x_n$, and a set of lines connecting the successive iterations of the map. The time sequence produced by the mapping is obtained by choosing a value of μ (in this case $\mu = 2$), plotting the corresponding quadratic curve, and repetitively generating subsequent points starting with some initial value (in this case, $x_0 = 0.2$). The first point, x_1, is found where the line $x_0 = 0.2$ meets the quadratic curve. The next step is easily determined by moving laterally to the $x_{n+1} = x_n$ diagonal. From the diagonal x_2 can be found by again going vertically to the quadratic curve. The process is repeated until x settles (in this case) to a steady state where $x_{n+1} = x_n$.

Such a 'fixed' point is obtained whenever the *magnitude of the slope* of the map where it intersects the diagonal is less than unity.

In Figure 4.1 the sequence $\{x_n\}$ reaches a fixed point. A little experimentation shows that this is apparently the case for all initial conditions x_0 when $\mu = 2$. If μ is increased to approximately 3.3 as in Figure 4.2 the situation changes. The quadratic curve is steeper and the magnitude of its slope $|f'(x)|$ is greater than 1 at the intersection. Therefore, the fixed point is unstable and, after an initial transient, x_n oscillates between two values so that $x_{n+2} = x_n$. This effect is reminiscent of the Poincaré map of the pendulum for values of the driving force around $g = 1.07$. Higher values of μ lead to further bifurcations and even to chaotic behavior. Figure 4.3 shows the situation for $\mu = 3.9$. The long-term behavior is such that the x_n are not limited to a few points but rather fill much of the original quadratic curve, and the behavior is chaotic. This behavior is reminiscent of the Poincaré sections for the pendulum in the chaotic region.

Rather than continuing to describe the behavior of the logistic map for individual values of μ, we present a more global view of the model through a bifurcation diagram, as shown in Figure 4.4, where μ varies smoothly from 2.9 to 4.0. In this diagram the map is iterated several hundred times at each of many intervening values of μ, with the first 100 values discarded to ensure that only the long-term behavior is

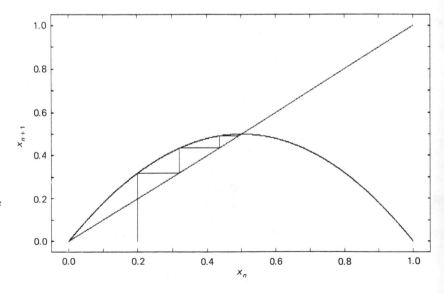

Fig. 4.1 Evolution of the logistic map for $\mu = 2$. The equilibrium value is $x = 0.5$.

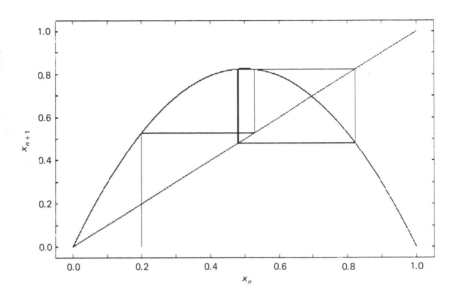

Fig. 4.2 The logistic map for $\mu = 3.3$ showing an oscillation between $x = 0.48$ and $x = 0.83$.

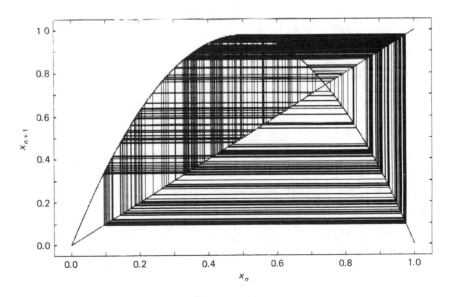

Fig. 4.3 Iteration of the logistic map for a chaotic state at $\mu = 3.9$.

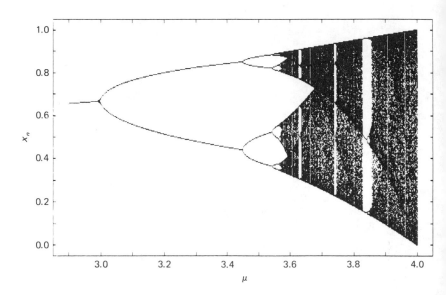

Fig. 4.4 Bifurcation diagram of the logistic map. Long-term values of x_n are plotted for $2.9 < \mu < 4$.

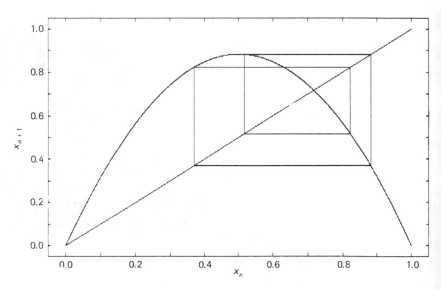

Fig. 4.5 A period-4 logistic map cycling between four values: $x = 0.37$, 0.52, 0.83, and 0.88.

plotted. The appearance of this diagram is similar to that of the pendulum bifurcation diagram (Figure 3.13), including regions where the behavior is chaotic and regions or *windows* of periodicity. We now focus on some general features of chaotic maps that are illustrated by the logistic map.

Period doubling

One important feature of the logistic map is the passage to chaos through a sequence of period doublings; the bifurcation where this doubling occurs is called a *pitchfork* bifurcation, because the local shape of the bifurcation diagram resembles a pitchfork. This *period doubling* effect is illustrated in Figure 4.5, which shows the long-term behavior of the map at $\mu = 3.53$. Two such bifurcations have occurred and $x_{n+4} = x_n$.

The period-doubling mechanism is one *route to chaos* that has been much studied as it is common in many dynamical systems, including the pendulum for g slightly greater than 1. The period-doubling route is particularly interesting because it may be characterized by certain universal numbers that do not depend (within certain limits) on the nature of the map (or ordinary differential equation). For example, the ratio of the spacings between consecutive values of μ at the bifurcations approaches a universal constant, called the Feigenbaum number after its discoverer. If the first bifurcation occurs at μ_1, the second at μ_2, and so forth, then this universal number is defined as (Feigenbaum, 1978)

$$\lim_{k \to \infty} \frac{\mu_k - \mu_{k-1}}{\mu_{k+1} - \mu_k} = \delta = 4.669\,201\,609\,102\,990\,9 \ldots.$$

This number can be roughly checked by careful scrutiny of the bifurcation diagram. Furthermore, it can be used to generate the sequence $\{\mu_k\}$, using the bifurcation diagram to select the first few values. Finally, it can be shown that an infinite number of bifurcations occur as $\mu = 3.569\,944 \ldots$ is approached.

The Feigenbaum number is a universal property of the period-doubling route to chaos for maps that have a quadratic maximum. *Universality* expresses the notion that certain properties of nonlinear maps are independent of the specific form of the map.

The periodic windows

The regions of chaotic behavior are interrupted by intervals of periodic behavior for $\mu > \mu_x = 3.569$. One of the largest of these *windows* occurs near $\mu = 3.83$, where a periodic orbit (a 3-cycle) occurs, as shown in Figure 4.6. The existence of this periodic behavior is evident from the shape of the third return map, for which

$$x_{n+3} = f(f(f(x_n))).$$

In Figure 4.7 two such maps are shown for two values of μ; (a) at the start of the window where $\mu = 3.8282$, and (b) inside the window before the period doubling cascade begins, where $\mu \approx 3.84$. Although these diagrams have very similar appearances there are some important differences.

At the left boundary of this window, the third order return map (Figure. 4.7(a)) shows three values of x where the curve is tangent to the diagonal line, $x_{n+3} = x_n$. These points are the cyclic steady state values of x which appear at the beginning of the window. Other initial values of x will be drawn to these fixed points since the shallow slopes of the curve near the fixed points lead to stability. This particular type of transition is called a *tangent bifurcation*.

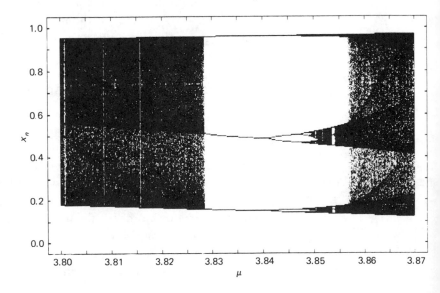

Fig. 4.6 Magnification of the bifurcation diagram in the region of the period-3 window.

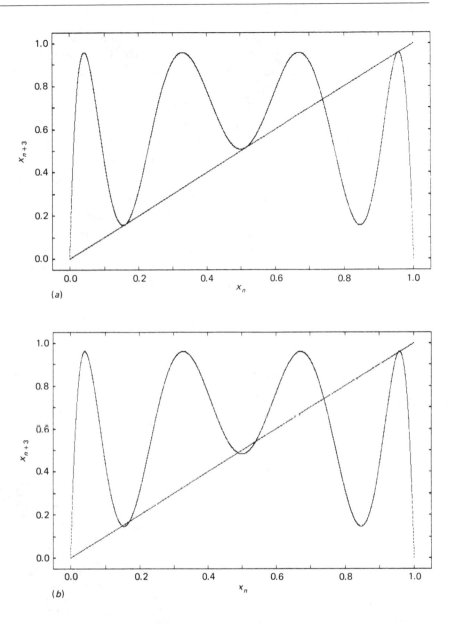

Fig. 4.7 (a) The map $x_{n+3} = f^3(x_n)$ at the onset of the period-3 window ($\mu = 3.8282$) showing three stable values of x_n at about 0.16, 0.51, and 0.95. An unstable point where $x_{n+3} = x_n$ appears at about $x_n \approx 0.76$. The origin of the term 'tangent bifurcation' is apparent. (b) The map $x_{n+3} = f^3(x_n)$ just inside the window at $\mu = 3.84$.

For slightly larger μ the bifurcation diagram continues to show the period-3 behavior. Figure 4.7(*b*) shows the behavior of the corresponding third order return map. The curve now crosses the diagonal at three pairs of values of *x*. (The isolated crossing is unstable.) The slopes of the curve in the neighborhood of three of the points (one from each close pair) are sufficiently steep that the map wanders away from these fixed points. On the other hand, the magnitudes of the slopes at the other three points are less than 1 so these points are attractors. Therefore the cyclic behavior, initiated by the tangent bifurcation, continues to be stable.

At $\mu = 3.842$ a subharmonic cascade to chaos occurs. The slopes of the third order maps near the previously stable values of *x* now become too steep for stability. For the first period doubling the sixth order map has six attractors; this process continues until chaotic bands form at $\mu \approx 3.85$. The resulting bands merge near $\mu \approx 3.857$ to form a continuum of values of *x*. This expansion of the chaotic regime and similar discrete changes in a chaotic attractor are sometimes called *crises* (Grebogi, Ott, and Yorke, 1987).

For values of μ just below the onset of the period-3 window, the third order return map is not quite tangent to the diagonal line. Therefore x_n can pass through the resulting narrow gaps and then go freely around the plane until it again becomes temporarily trapped in

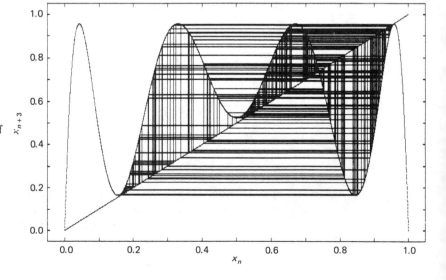

Fig. 4.8 An illustration of 'type I' intermittency as the trajectory squeezes through the gap between the map and the tangent line. During the passage through the gap, *x* changes very slowly.

a narrow gap as shown in Figure 4.8. While it is in the gap, x_n is nearly fixed. If we think of the map as a Poincaré section for a differential equation representing a physical system such as the pendulum, the physical variable would show nearly periodic motion with occasionally irregular bursts. This common type of chaotic motion is called *type I intermittency* and occurs when a dynamical system is close to a tangent bifurcation if there is a mechanism for intermittent return to the narrow gap.

Lyapunov exponent

The Lyapunov exponent of a map (named after A.M. Lyapunov, 1857–1918, a Russian mathematician) may be used to obtain a measure of the sensitive dependence upon initial conditions that is characteristic of chaotic behavior. This exponent (often written as λ) may be readily computed for a one-dimensional map such as the logistic map. If a system is allowed to evolve from two slightly differing initial states, x and $x + \varepsilon$, then after n iterations their divergence may be characterized approximately as

$$\varepsilon(n) \approx \varepsilon e^{\lambda n},$$

where the Lyapunov exponent λ gives the average rate of divergence. (The average must be taken over many 'initial conditions' spread over the trajectory.) If λ is negative, slightly separated trajectories converge and the evolution is not chaotic. If λ is positive, nearby trajectories diverge; the evolution is sensitive to initial conditions and therefore chaotic.

Consider a specific one-dimensional map given by $x_{n+1} = f(x_n)$. The difference between two initially nearby states after the nth step is written as

$$f^n(x + \varepsilon) - f^n(x) \approx \varepsilon e^{n\lambda},$$

or

$$\log_e \left[\frac{f^n(x + \varepsilon) - f^n(x)}{\varepsilon} \right] \approx n\lambda.$$

For small ε, this expression becomes

$$\lambda \approx \frac{1}{n} \log_e \left| \frac{\mathrm{d} f^n}{\mathrm{d} x} \right|.$$

Finally, we use the chain rule for the derivative of the nth iterate and take the limit as n tends to infinity to obtain

$$\lambda = \lim_{n \to \infty} \frac{1}{n} \sum_{i=0}^{n-1} \log_e |f'(x_i)|.$$

Therefore the Lyapunov exponent gives the stretching rate per iteration, averaged over the trajectory. In Figure 4.9, the Lyapunov exponent is plotted as a function of the parameter μ. The sign of λ correlates very well with the behavior of the system as shown in the bifurcation diagram, Figure 4.4. Beyond $\mu_\infty = 3.56$, the regions of periodic behavior correspond to the intervals in which $\lambda < 0$.

For n-dimensional maps there are n Lyapunov exponents, since stretching can occur for each axis. An n-dimensional initial volume develops, on average, as

$$V = V_0 e^{(\lambda_1 + \lambda_2 + \cdots + \lambda_n)n}.$$

For a dissipative system the sum of the exponents must be negative. If the system is chaotic then at least one of the exponents is positive. (See Problem 14 for a soluble two-dimensional map.)

Lyapunov exponents are also defined for continuous time dynamical systems such as the pendulum. An initial n-dimensional volume

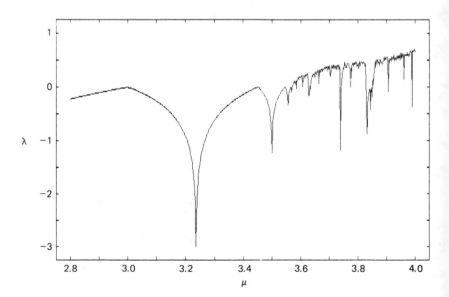

Fig. 4.9 Lyapunov exponent λ versus μ for the logistic map. Sensitivity to initial conditions occurs where the exponent is positive.

of phase space develops on average as

$$V = V_0 e^{(\lambda_1 + \lambda_2 + \cdots + \lambda_n)t}.$$

For the driven pendulum there are three Lyapunov exponents corresponding to the three dimensions of the phase space (θ, ω, ϕ). Since the orbits are solutions to a set of ordinary differential equations, the calculation of Lyapunov exponents is less straightforward than for maps. On a chaotic attractor such as that of the pendulum at $g = 1.5$, the directions of divergence and contraction are locally defined, and the calculation must constantly adjust for this condition. Despite this difficulty, computer algorithms have been developed for calculating Lyapunov exponents both from differential equations and from experimental data; the Lyapunov exponents of the pendulum are discussed further in Chapter 5.

Entropy

The complex appearance of the various graphical representations of chaotic behavior naturally leads to the question of the relationship between statistical mechanics and chaos. One way to connect these phenomena is to apply the concept of *entropy* to a chaotic system, comparing the result to an associated statistical system. This comparison is readily done with the logistic map.

Consider a hypothetical statistical system for which the outcome of a certain measurement must be located on the unit interval. If the line is subdivided into N subintervals, we can associate a probability p_i with the ith subinterval containing a particular range of possible outcomes. The entropy of the system is then defined as

$$S = - \sum_{i=1}^{N} p_i \log_e p_i.$$

This quantity may be interpreted as a measure of the amount of disorder in the system or as the information necessary to specify the state of the system. If the subintervals are equally probable so that $p_i = 1/N$ for all i, then the entropy reduces to $S = \log_e N$, which can be shown to be its maximum value. (See Problem 5.) Conversely, if the outcome is known to be in a particular subinterval, then $S = 0$, the minimum value. When $S = \log_e N$, the amount of further information needed to specify the result of a measurement is at a maximum. On the other hand when $S=0$ no further information is required. (See, for example, Baierlein (1971), for a discussion of the interpretation of entropy as 'missing' information.)

We now apply this formulation to the logistic map by establishing N 'bins' or subintervals of the unit interval into which the values of x_n may fall. In the nonchaotic state the x_n will fall in relatively few of the bins, and the entropy is low. But in the chaotic state, the entropy is higher, and if the frequencies of occurrence are equal, then it approaches $\log_e N$. Figure 4.10 shows the results of applying the entropy concept to the logistic map. As expected, the entropy generally increases with μ, except for the downward spikes in the windows of periodic behavior. The entropy does not quite get to $\log_e N$ until $\mu=4$, since the distribution of x_n does not span the whole unit interval evenly for $\mu<4$. This feature may be observed in the bifurcation diagram. For $\mu=4$, the entropy is similar to that of a random process with a uniform probability distribution. Nevertheless, short-term correlations *do* exist for chaotic motion, but not for the idealized random system.

For dynamical systems such as the pendulum, information changes in time. It turns out that the average temporal rate of information change can be related both to the Lyapunov exponents and to the

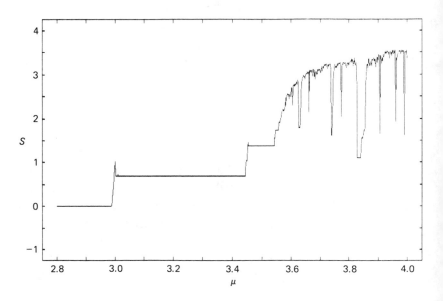

Fig. 4.10 Entropy S as a function of μ for the logistic map. The maximum entropy corresponding to equal probability for each of the 40 cells is 3.6888.

fractal dimension of the attractor. These subjects and their relationships are discussed in Chapter 5.

Stretching and folding

The logistic map also provides some insight into the stretching and folding mechanism that is necessary to keep chaotic trajectories within a finite volume of phase space, despite the exponential divergence of neighboring states. For the logistic map the stretching (divergence of neighboring trajectories) and folding (confinement to the bounded space) can be demonstrated fairly easily, by reference to Figure 4.11.

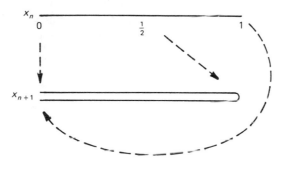

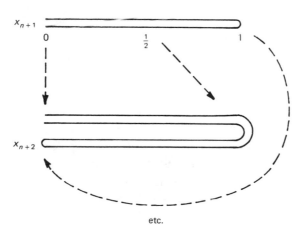

Fig. 4.11 The stretching and folding property of the logistic map for $\mu = 4$.

For $\mu = 4$ the logistic map has a maximum value of 1 at $x_n = \frac{1}{2}$ and the values of $x_n \in (0, \frac{1}{2})$ map to $x_{n+1} \in (0,1)$. Similarly the values of $x_n \in (\frac{1}{2}, 1)$ map to $x_{n+1} \in (0,1)$, but in the reverse order. Therefore both intervals of x_n are stretched by a factor of 2, but because the order of the mappings is opposite, the second stretched interval is folded onto the first stretched interval. The figure illustrates a few cycles of the mechanism. The process resembles the one used to make taffy candy or knead dough for bread.

The stretching and folding process illustrates another important feature of chaotic systems that was implied in our discussion of entropy, namely the loss of information about the initial conditions of a system as time or iteration number increases. Mathematically this arises from the *noninvertibility* of the map $f(x_n, \mu)$. That is, it is always possible to predict x_{n+1} from x_n but there is ambiguity in trying to retrodict x_n from x_{n+1}. (One finds the same noninvertibility with elementary functions such as $y = \sin x$, $y = x^2$, and so forth. The inverse functions can be defined only by limiting the original domains.) It turns out that a necessary condition for any one-dimensional map to exhibit chaotic behavior is that it be noninvertible.

The circle map

The logistic model illustrates many characteristics of chaotic dynamics, such as bifurcations, period doubling, intermittency, sensitivity to initial conditions, and the stretching and folding process. However, some important features of the pendulum, especially the phenomenon of 'phase locking', require a two-parameter map for their explanation. Phase locking is said to occur when the ratio of the frequency of the pendulum to that of the forcing becomes locked at the ratio p/q of two integers, over some finite domain of parameter values (D'Humieres *et al.*, 1982). A similar phenomenon was observed by Christian Huygens in the seventeenth century: the synchronization of two clocks on the same wall. The common attachment to the same wall must have provided a coupling of the clocks to each other. (This phenomenon is mentioned in Bak (1986).)

The pendulum's Poincaré section may be modeled as a two-dimensional (but unknown) map:

$$\theta_{n+1} = G_1(\theta_n, \omega_n)$$
$$\omega_{n+1} = G_2(\theta_n, \omega_n).$$

If ω_n is a function only of θ_n after the initial transients have died away, then $\omega_n = f(\theta_n)$, and the two-dimensional Poincaré map reduces to a one-dimensional map:

$$\theta_{n+1} = G_1(\theta_n, f(\theta_n))$$

or

$$\theta_{n+1} = F(\theta_n).$$

This map may be regarded as a mapping of the circle to itself. It is one-dimensional, with an angular coordinate $\theta_n \in [0,1]$ and periodic boundary conditions (corresponding to the pendulum angular coordinate, $\theta \in [0,2\pi]$).

For a certain range of forcing amplitudes and frequencies, a circle map may be a reasonable approximation to the driven pendulum. The difference equation of a particularly useful circle map known as the *standard map* is

$$\theta_{n+1} = \theta_n + \Omega - (K/2\pi)\sin(2\pi\theta_n) \qquad \text{mod } 1.$$

There are two parameters (Ω, K) for the standard map, in contrast to the single parameter μ for the logistic map. The parameter Ω is the rotation frequency ('winding number') in the absence of nonlinearity, whereas K gives the strength of the nonlinear coupling of the oscillator to the forcing. This nonlinear coupling can modify the angular change per iteration. (A numerical justification of the connection between the standard map and the pendulum over a range of parameters is given in Jensen, Bak, and Bohr (1984).)

To obtain a sense of the behavior of the standard map, we first omit the nonlinear term by setting $K = 0$. Then the map reduces to

$$\theta_{n+1} = \theta_n + \Omega,$$

which is illustrated in Figure 4.12 for the case $\Omega = 0.4$. After five iterations θ returns to its initial value $\theta_0 = 0.3$, having made two revolutions. The *winding number*, W, is $\frac{2}{5}$, and it is just equal to Ω. If the winding number is a rational number, p/q, then the map is cyclic or *periodic*. If the winding number is irrational then θ does not return exactly to its initial value and the motion is termed *quasiperiodic*.

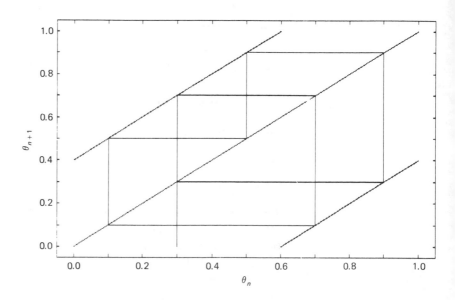

Fig. 4.12 The linearized circle map for a rational winding number of 0.4, using periodic boundary conditions on θ. The diagonal line represents $\theta_{n+1} = \theta_n$.

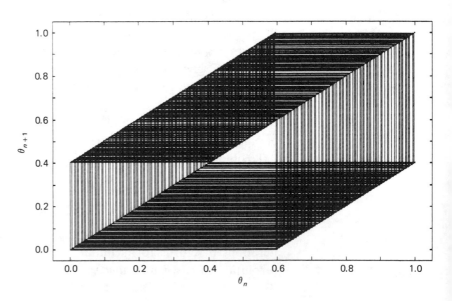

Fig. 4.13 The linearized circle map for an irrational winding number, 0.404004 . . . .

Figure 4.13 illustrates quasiperiodic motion with $\Omega = 0.404\,004\ldots$ (irrational) for 200 iterations. The angle comes arbitrarily close to any particular value if n is sufficiently large. Mode locking occurs when the nonlinear term is added; this keeps the motion periodic even when Ω is irrational. In Figure 4.14, for example, $K = 0.95$ and $\Omega = 0.404\,004\ldots$ as before. However, the motion repeats every five iterations. The winding number measures the average phase change per iteration. For $K \neq 0$, it is not equal to Ω and is defined generally as

$$W = \lim_{n \to \infty} \left(\frac{\theta_n - \theta_0}{n} \right).$$

The nonlinear term obviously changes the shape of the function representing the map. Note that at $K = 0.95$ the map is still invertible. The widths in Ω of the various mode-locked regions where the winding number is fixed increase with K, as shown in Figure 4.15. The resulting 'Arnold tongues' are named after the Russian mathematician who discovered this structure (Arnold, 1965).

There are an infinite number of phase-locked intervals. There are also an infinite number of irrational winding numbers. As Ω varies at fixed K, the map displays both periodic and quasiperiodic motion. But as K approaches 1, the rational intervals increase in size. At $K = 1$ the set of rational intervals is a fractal. Figure 4.16(a) shows the rational winding numbers as plateaus in a plot of W versus Ω. If the figure is magnified (Figure 4.16(b)), more plateaus become evident, and the curve shows repetition of the same patterns at the new magnification. Such a curve is said to be *self-similar*. This structure is called the *Devil's staircase*. (For a discussion of the Devil's staircase and some applications, see Bak (1986).)

Beyond the $K = 1$ critical value, the phase-locked motions overlap; this implies that several different periodic oscillations can occur for given (K,Ω) depending on initial conditions. The graph of W versus Ω ceases to be monotonic. The map develops local maxima and minima and therefore becomes noninvertible for $K > 1$, a necessary condition for chaotic behavior, as we also noted for the logistic map. A noninvertible case of the standard map is shown in Figure 4.17. Chaos is, in fact, observed for some values of Ω.

Several routes to chaos occur for the standard map. In Figure 4.18 we illustrate three representative paths through the Arnold tongues of the (Ω,K) parameter space. Path (a) shows the system in a state where

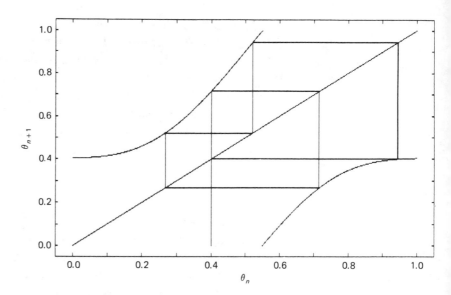

Fig. 4.14 The standard map for $K=0.95$ and $\Omega=0.404\,004\,\ldots$. The nonlinear coupling produces a phase-locked state.

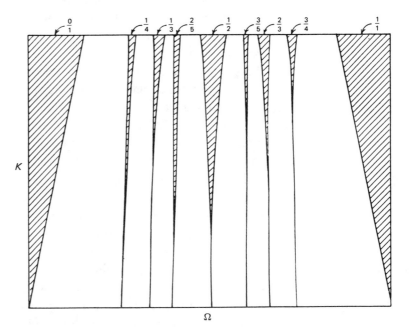

Fig. 4.15 Parameter space of the standard map as a function of K and Ω. The motion is periodic (rational winding number) inside the 'Arnold tongues', and quasiperiodic outside them. At $K=1$ (top of figure) only rational winding numbers are available and for $K>1$ the tongues overlap. (Only a few of the 'tongues' are illustrated.)

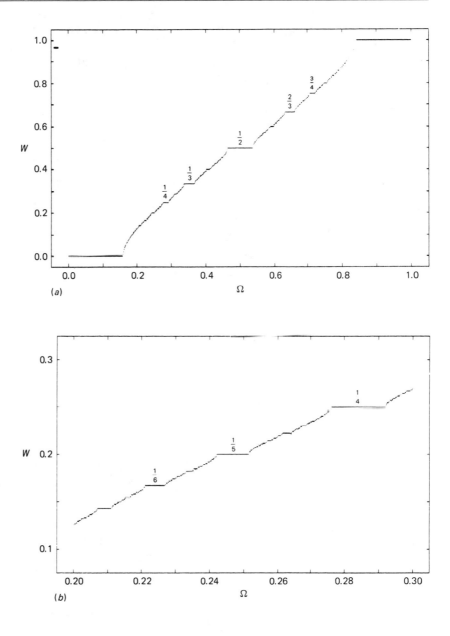

Fig. 4.16 (*a*) The 'Devil's staircase' generated by the rational winding numbers of the standard map. $K = 1$. (*b*) A magnification of part of the staircase. The low-order phase-locked tongues are labeled by their winding numbers.

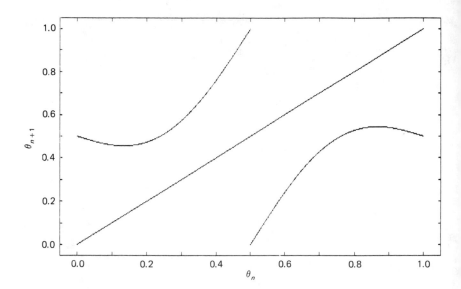

Fig. 4.17 The standard map becomes noninvertible for $K > 1$ (here $\Omega = 0.5$ and $K = 1$).

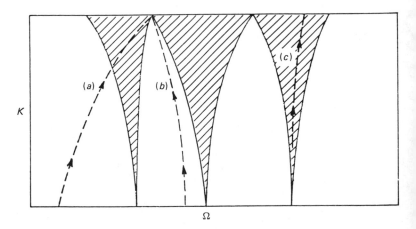

Fig. 4.18 Various possible routes to chaos: (*a*) quasiperiodic→phase locking→chaos; (*b*) quasiperiodic→chaos; (*c*) periodic→chaos.

the winding number is irrational and the behavior is quasiperiodic. The system continues in the quasiperiodic state until it reaches the junction of two phase-locking modes at $K = 1$ and becomes chaotic. Path (b) shows the system as initially quasiperiodic but then passing into a mode-locking regime and eventually becoming chaotic. Path (c) shows the system starting in a mode-locked state, traveling toward the critical line, $K = 1$, but then continuing to a larger value of K, beyond which a period-doubling cascade to chaos begins. The prominence of the period-doubling route to chaos is consistent with the existence of a quadratic maximum (Figure 4.17). This aspect of the circle map is similar to the logistic map.

The logistic and circle maps provide many valuable insights into chaotic dynamics. As we discuss in the final section of this chapter, many of the concepts developed from these one-dimensional maps apply to the driven pendulum. As a final model for chaotic behavior we consider a two-dimensional map.

The horseshoe map

In our discussion of the logistic map we saw that the interval $(0,1)$ is stretched and then folded back upon itself. The stretching and folding phenomenon is a primary mechanism for allowing sensitivity to initial conditions in a finite-sized phase space. The horseshoe map introduced by Smale (1963) is a two-dimensional mapping that illustrates the stretching and folding action. It has been shown to be embedded in the dynamics of the pendulum for some parameter choices (Gwinn and Westervelt, 1986).

The horseshoe map consists of the sequence of operations shown in Figure 4.19. First consider a map f which acts upon the unit square, and consists of (a) an expansion in the y direction by a factor $\mu > 2$, (b) a contraction in the x direction by a factor $\lambda \in (0, \frac{1}{2})$, and (c) a folding, as illustrated in Figure 4.19. The transformed set $f(S)$ is then intersected with the original set S so that the map is now confined to a subset of the original unit square. If the entire sequence of operations is repeated, then four stripes appear from the original two, and so on. Repetition of the process n times leads to 2^n stripes, and a cut across the stripes would, in the limit of large n, lead to a fractal (see Chapter 5).

Horseshoe configurations occur in the phase space of dynamical systems where there are regions of strong contraction and expansion. For example, we recall from the initial discussion (Chapter 2) of the pendulum phase plane that there are saddle points at $\theta = \pm \pi$ and $\omega = 0$. Near these saddle points, trajectories approach most rapidly along certain 'stable' directions, and depart most rapidly along other 'unstable' directions, as shown in Figure 2.13. Along these directions, the Lyapunov exponents are negative and positive, respectively. Alternatively stated, tangent vectors along the stable directions are contracting, and tangent vectors along the unstable directions are expanding. Any region of phase space where these two types of

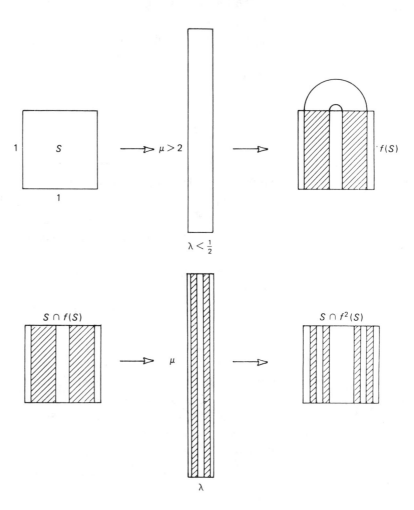

Fig. 4.19 The construction of the horseshoe map for two iterations.

behavior are in close proximity may exhibit stretching and folding.

Trajectories containing both types of behavior develop in a complex way. For example, see Moon (1987), Chapter 5. This may be explained with reference to Figure 4.20, which shows the phase plane for the damped, undriven pendulum as in Figure 2.13, but with the *stable and unstable manifolds* W^s and W^u of the saddle points labeled, and the two basins of attractions shaded differently. The manifolds W^s and W^u are simply the trajectories that approach and depart most quickly from the unstable equilibrium. If the pendulum is now *driven periodically* but weakly, the same diagram may be regarded as a Poincaré section of the three-dimensional phase space, except that the lines should be regarded as a sequence of dots corresponding to successive passages of the trajectories through the Poincaré plane.

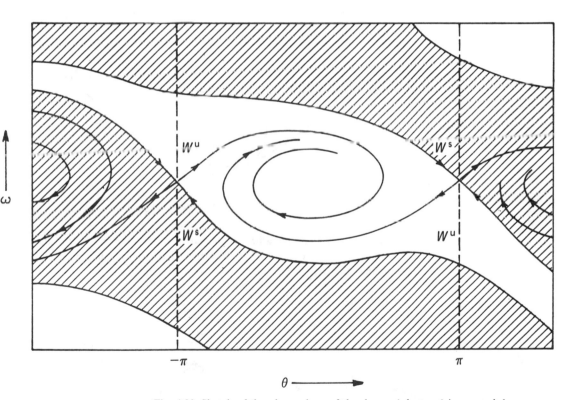

Fig. 4.20 Sketch of the phase plane of the damped, but undriven pendulum, showing the stable and unstable manifolds of the saddle points. Shaded and unshaded regions correspond to distinct basins of attraction.

If the pendulum is driven more strongly, the unstable manifold from the saddle at π and the stable manifold from the saddle at $-\pi$ may approach each other and touch as shown in Figure 4.21(a), or even cross at the point I_1 in Figure 4.21(b). (The actual trajectories do not cross of course, but the stable and unstable manifolds in the Poincaré section can cross.) Now comes the surprise. Each crossing is mapped into another one closer to the saddle point, leading to an infinite number of intersections I_2, I_3, and so forth. The resulting

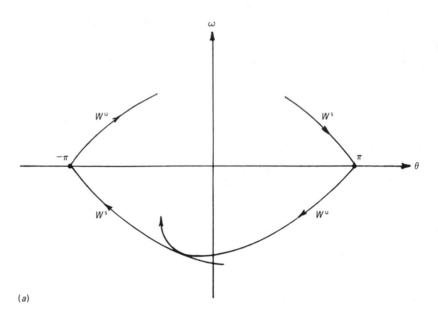

(a)

Fig. 4.21 The formation of a heteroclinic tangle in the Poincaré map of the pendulum. (a) The unstable and stable orbits barely touch, signaling the beginning of chaos. (b) The tangle forms with an infinite number of intersections $I_1, I_2, \ldots$. Two nearby points may be mapped far apart, yielding chaos. (c) Additional detail (see text).

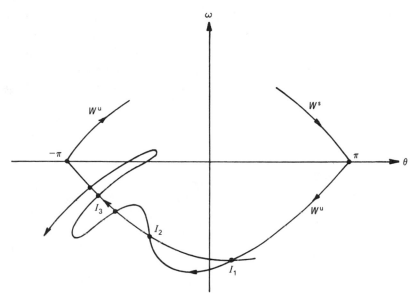

(b)

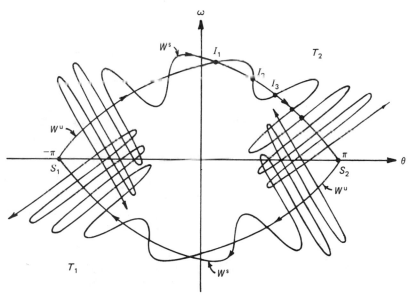

(c)

configuration is called a *heteroclinic tangle*. (If W^s and W^u come from the same fixed point, the configuration is known as *homoclinic*.)

Because of the strong bending of the manifolds near the saddle point, a small rectangular section of the plane near I_1 will suffer stretching and folding much like that of the horseshoe map. In fact, that distorted rectangle is topologically equivalent to (can be smoothly deformed into) the iterated Smale horseshoe (Abraham and Shaw, 1984). As a result, two points that are initially close together will be found far apart after a few iterations. Therefore, chaos is a natural consequence of a heteroclinic tangle.

The actual situation is even more complicated than Figure 4.21(*b*) suggests. Let us label the first tangle near saddle S_1 as T_1 (see Figure 4.21(*c*)). Clearly there must be a second tangle T_2 near the saddle S_2 at π, since the geometry there is the same as that near S_1. But where did the first intersection I_1 in that tangle come from? It must have resulted from an earlier iteration. Going *backward* in time takes I_1 back through an infinite sequence of intersections to the neighborhood of S_1. This implies that the stable and unstable manifolds from S_2 must cross each other an infinite number of times near S_1. Thus, the geometry of the pendulum (as visualized in the Poincaré plane) is infinitely complex, and the essential character of that complexity is contained in the horseshoe map.

Applications to simulations of the pendulum

The logistic map, the standard map, and the horseshoe map illustrate the kinds of phenomena that are important aspects of the motion of the driven pendulum. Though we have alluded briefly to connections between the driven pendulum and the maps, we now discuss several of these connections in greater detail.

(*i*) *Period doubling.* The logistic map illustrates the period-doubling route to chaos. Reference to the bifurcation diagram of Figure 4.22 provides evidence of similar behavior for the pendulum. A pair of period-doubling cascades begins at $g \approx 1.07$ (preceded by symmetry breaking at $g = 1.0$, where the angle exceeds π). An examination of the data of Figure 4.22 (*b*) at greater magnification of g and ω than in Figure 4.22 (*a*) – shows period doubling at $g = 1.066$, $g = 1.077$, and

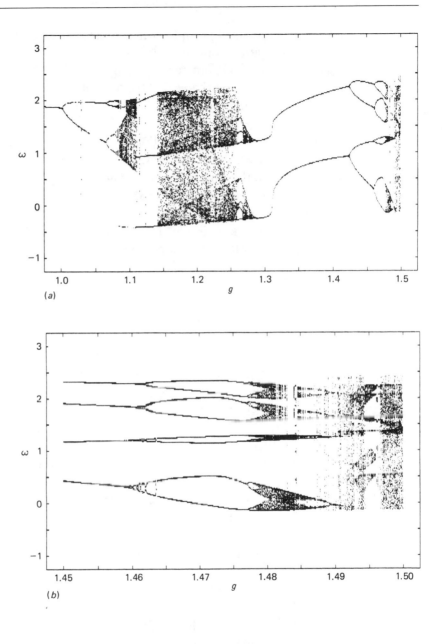

Fig. 4.22 Bifurcation diagrams for the pendulum, indicating various dynamical regimes. The diagrams are generated by following the long-term behavior of two initial points (θ_0, ω_0), one each from the positive $\langle\omega\rangle$ and negative $\langle\omega\rangle$ basins of attraction. (b) Magnification of part of (a).

$g \approx 1.080$, with further bifurcations unresolved. Using these data the ratios of the changes in g can be estimated and compared with the Feigenbaum number, $4.669\ldots$. For this sequence the result is 4 ± 1. It is remarkable that the behavior of the logistic map is manifested (to within the computational accuracy) in the more complex pendulum.

(ii) *Phase locking.* Phase locking of the pendulum is evident when the average angular velocity is some rational multiple (usually low order) of the angular forcing frequency ω_D. This condition may be specified in the following way. If the pendulum is phase locked at a ratio p/q, then for q drive periods the angle difference is $\theta(t+qT)-\theta(t)=2\pi p$, where T is the drive period, $2\pi/\omega_D$. Then the average value of $\omega = d\theta/dt$ over q periods is

$$\langle \omega \rangle = (1/qT) \int_t^{t+qT} \omega \, dt = (p/q)\omega_D.$$

Measurement or computation of the average angular velocity is a useful tool for analysis of the pendulum motion. A graph of $\langle \omega \rangle$ versus g as shown in Figure 4.23 should reveal phase-locked motion. This figure complements the bifurcation diagrams of Figure 4.22. Two sets of initial conditions were chosen, one from each basin of attraction at $g = 1.45$, to illustrate positive and negative rotary modes. The plateaus of $\langle \omega \rangle$ are indicative of phase locking; they correspond to the periodic intervals of the bifurcation diagram. The regions in which $\langle \omega \rangle$ varies erratically correspond to the chaotic state (Gwinn and Westervelt, 1986).

Another approach to the study of phase locking is direct examination of the winding number for a range of forcing amplitudes, g. In the diagrams of Figure 4.24 the winding number is shown for two ranges of g. It is defined for the pendulum as

$$W = \lim_{n \to \infty} \left(\frac{\theta_n - \theta_0}{2\pi n} \right),$$

where n is the number of drive cycles. As in the previous diagrams initial values were chosen from the two basins of attraction to show positive and negative angular velocities. For each value of g an initial motion corresponding to 50 drive cycles is discarded, and the next 30 cycles ($n = 30$) of angular displacement are used to obtain W. The features of these diagrams are essentially the same as those of the

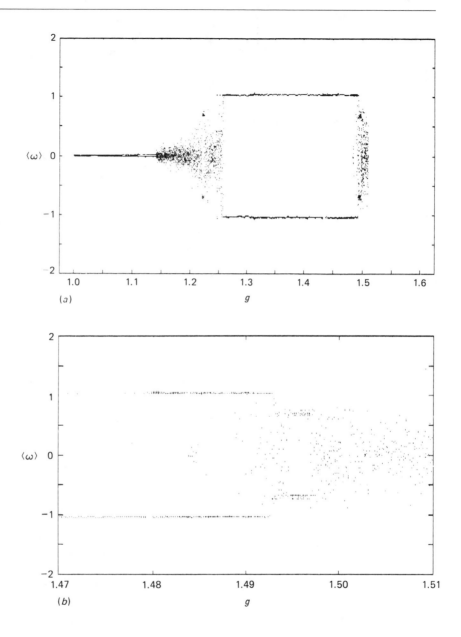

Fig. 4.23 Phase locking of the pendulum as revealed by the average angular velocity $\langle\omega\rangle$ (in units of ω_D) as a function of the driving force amplitude g. Sets of initial coordinates (θ_0, ω_0) were chosen from the two basins of attraction. (b) Magnification of part of (a).

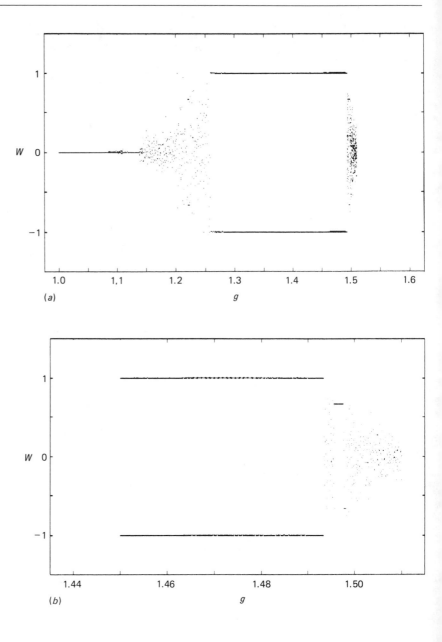

Fig. 4.24 Phase locking of the pendulum as revealed by the winding number W as a function of driving amplitude g. Sets of initial coordinates were chosen from the two basins of attraction. (b) Magnification of part of (a).

Table 4.1. *Correlation of dynamical behavior with winding number. Behavior designated 'chaotic(ma)' indicates multiple chaotic attractors. $q = 2$ and $\omega_D = \frac{2}{3}$*

Range	Type of behavior	Winding number
$g < 1.085$	periodic	0
$1.085 < g < 1.11$	chaotic(ma)	~ 0
$1.11 < g < 1.14$	periodic	0
$1.14 < g 1.22$	chaotic	scattered
$g \sim 1.22$	periodic	$\pm \frac{2}{3}$
$1.22 < g < 1.26$	chaotic	scattered
$1.26 < g < 1.28$	chaotic(ma)	$\sim \pm 1$
$1.28 < g < 1.475$	periodic	± 1
$1.475 < g < 1.485$	chaotic(ma)	$\sim \pm 1$
$1.485 < g < 1.493$	periodic	± 1
$1.493 < g < 1.495$	chaotic	scattered
$1.495 < g < 1.497$	periodic	$\pm \frac{2}{3}$
$g > 1.497$	chaotic	scattered

graphs of $\langle \omega \rangle$ versus g, since both measure the average rotation rate, using different computational schemes.

The graph of W versus g shows two types of behavior; (*a*) phase locking with constant W for periodic pendulum states, and (*b*) scattered values of W for chaotic states. The various types of behavior detected as g increases, and the corresponding values of W, are shown in Table 4.1.

In this chapter we have demonstrated that discrete mappings can give insight into the complex behavior of the driven pendulum. In the next chapter we examine various aspects of the fractal geometry associated with chaotic motion.

Problems

1. Use the listing LOGISTIC MAP in Appendix B or the option LOGISTIC MAP from the CHAOS menu to study the logistic return map. Try different values of the parameter μ and different initial values x_0.

2. Use one of the programs suggested in Problem 1 to study regions where period doubling occurs. First look at the appropriate first order return map and then generate higher order return maps that correspond to the degree of period doubling.

3. Using one of the programs suggested in Problem 1, generate a bifurcation diagram for the logistic map. Expand the scale of μ in order to magnify certain regions of the diagram. In particular, expand the scale in a chaotic region and note that windows of periodic behavior are more evident at the higher magnifications.

4. Expand the scale of μ for a bifurcation diagram in the region of period doubling. Try to observe many bifurcations and thereby approximately verify the Feigenbaum number.

5. For the second bifurcation of the logistic map, the entropy, as defined in the text, is constant over that region of μ. What does this fact imply about the distribution of points over the four possible values of x_n for that range of μ? Now assume that the values of x_n are tossed with equal probability into each of four bins (out of a total of 40 bins). What is the entropy of this situation? Compare your answer with that of Figure 4.10, and suggest an interpretation of the diagram in that particular region. Using the method of Lagrange multipliers or otherwise, prove that the entropy function is a maximum when $p_i = 1/N$ for all values of i.

6. Another map which shares many properties of the logistic map is the *tent* map:

$$x_{n+1} = 2\beta x_n \qquad \text{for } 0 < x < \tfrac{1}{2} : 0 < \beta < 1$$
$$x_{n+1} = 2\beta(1 - x_n) \text{ for } \tfrac{1}{2} < x < 1$$

Use either the TENT MAP option from the CHAOS menu or your own modification of the listing LOGISTIC MAP in Appendix B to generate some mappings and bifurcation diagrams of the tent map.

7. Use one of the programs suggested in Problem 6 to generate a plot of the Lyapunov exponent versus β for the tent map. Prove analytically that the Lyapunov exponent is $\log_e(2\beta)$. Note that the exponent becomes positive as β passes through 0.5, the initial point of chaotic behavior.

8. In the chaotic region of the tent map it is possible to estimate how many iterations are necessary before knowledge of the x coordinate (with an initial uncertainty) is lost. If the uncertainty in the coordinate after the nth iteration is ε_n then the uncertainty after the $n+1$ iteration is

$$\varepsilon_{n+1} = \varepsilon_n e^{\log_e 2\beta}$$

(This expression uses the Lyapunov exponent from Problem 7.) If the initial uncertainty is ε how many steps does it require to have an uncertainty equal to 1? (Answer: $n = \log_2(1/\varepsilon)$)

9. Show that the logistic map (with $\mu = 4$) with the variable x_n may be transformed to the tent map with the variable y_n by the coordinate transformation: $y_n = (2/\pi)\sin^{-1}(x_n^{\frac{1}{2}})$.

10. Using either the listing CIRCLE MAP in Appendix B or the option CIRCLE MAP from the CHAOS menu generate the standard map using various values of K and Ω. Determine the differing effects of each of these parameters on the shape of the map. For what value of K does the map become noninvertible?

11. Using one of the programs suggested in Problem 10 generate several versions of the Devil's staircase. By appropriate scaling of the coordinates examine the staircase at various magnifications.

12. The phenomenon of mode locking in the driven pendulum can be examined by considering a modified version of the bifurcation diagram. Instead of keeping ω_D constant and varying g, reverse the operation and let ω_D be the independent variable, for constant g. (This requires the appropriate modification of the bifurcation program.) Try $g = 1.46$ for example and let ω_D vary from 0 to 1. You should observe that in the region where $\omega_D \approx p/q$ for small integer values of p and q, the pendulum locks onto a periodic motion.

13. Horseshoes can be generated in a variety of ways. One example is the baker's transformation:

$$x_{n+1} = 2x_n, \bmod 1$$

$$y_{n+1} = \begin{cases} ay_n & \text{for } 0 \le x_n < \frac{1}{2}, \\ \frac{1}{2} + ay_n & \text{for } \frac{1}{2} \le x_n \le 1 \end{cases}$$

for $a \le \frac{1}{2}$. Consider the nondissipative case where $a = \frac{1}{2}$, and think about the map as involving several steps. First stretch the x direction by $x_{n+1} = 2x_n$. Then compress the y direction by using $y_{n+1} = \frac{1}{2}y_n$. Then cut the picture vertically in half by applying the mod 1 operation to the original x_n transformation. Finally, place the right block on top of the left block by adding $\frac{1}{2}$ to all y_{n+1} for which x_{n+1} is greater than $\frac{1}{2}$. Write a computer program that will transform a rectangular block of initial coordinates inside the unit square according to the baker's transformation, using $a < \frac{1}{2}$. If the program runs many times the result should be a fractal (see Chapter 5).

14. The baker's transformation is two-dimensional and therefore has two Lyapunov exponents, λ_x and λ_y. Determine these exponents by considering the stretching process in the x direction and the contraction in the y direction.   (Answer: $\lambda_x = \log_e 2$, $\lambda_y = \log_e a$)

15. Another map that can produce horseshoes is the Hénon map:
$$x_{n+1} = 1 - ax_n^2 + y_n$$
$$y_{n+1} = bx_n.$$

If $b = 1$ then the map preserves areas, and if $|b| < 1$ then the map is dissipative. Write a program to transform an initial block of (x,y) coordinates according to the Hénon map.

16. Beyond the symmetry breaking of the pendulum at $g \approx 1$ there is a sequence of period doubling. Use the BIFURCATION program through a range of g that covers the doubling region and check the ratio:
$$\frac{g_{n+1} - g_n}{g_{n+2} - g_{n+1}},$$

where g_n is the value of g at the nth bifurcation. Does this ratio appear to approach the Feigenbaum ratio? Follow a similar procedure for the doubling in the region $g > 1.4$.

CHAPTER FIVE

The characterization of chaotic attractors

Many of the geometric structures generated by chaotic maps or differential dynamical systems are extremely complex. For example, the chaotic attractor of the pendulum (Figure 3.3(d)) in its three-dimensional phase space typically consists of an infinite number of infinitely thin layers. Its Poincaré section (Figure 3.5) reveals this structure clearly. These sets are called *fractals*. In this chapter we discuss fractals, their dimensions, and the relation between the fractal character of chaotic attractors and the underlying dynamics.

An elementary example of a fractal is the *Cantor set*. This set is a prototype of complex geometric structures in much the same way that the logistic map is a prototype for chaotic dynamical systems. The Cantor set is generated by iteration of a single operation on a line of unit length, as shown in Figure 5.1. The operation consists of

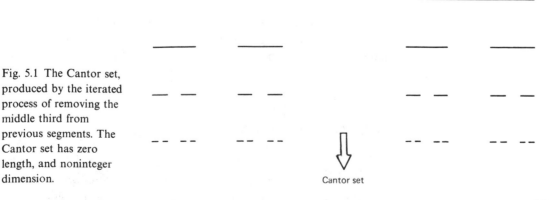

Fig. 5.1 The Cantor set, produced by the iterated process of removing the middle third from previous segments. The Cantor set has zero length, and noninteger dimension.

Cantor set

removing the middle third from each line segment of the previous set. As the number of iterations increases, the number of separate pieces tends to infinity, but the length of each one approaches zero. Furthermore, if the set is examined under high magnification, its structure is essentially indistinguishable from the unmagnified version. This property of invariance under a change of scale is called *self-similarity* and is common to many, although not all, fractals. (Within the resolution of the numerical simulation, the pendulum's Poincaré section of Figure 3.6 appears to exhibit this property.)

In contrast to a line with its infinite number of points and finite length, the Cantor set has an infinite number of points but zero length. Therefore it seems plausible that the dimension of the Cantor set should be less than 1 but greater than zero, the dimension of a finite set of points. The possibility of noninteger dimension requires a more sophisticated concept of dimension than that associated with lines, surfaces, and solids.

Dimension

There are many ways to define the dimension, $d(A)$, of a set A. (The particular set will often be omitted in the notation.) One approach is the *capacity* dimension, d_C.

Consider a one-dimensional figure such as a straight line or curve of length L, as shown in Figure 5.2(a). This line can be 'covered' by $N(\varepsilon)$ one-dimensional boxes of size ε on a side. If L is the length of the line then

$$N(\varepsilon) = L(1/\varepsilon).$$

Similarly, a two-dimensional square of side L can be covered by $N(\varepsilon) = L^2(1/\varepsilon)^2$ boxes as shown in Figure 5.2(b). For a three-dimensional cube the exponents would be 3, and so on for higher dimensions. In general,

$$N(\varepsilon) = L^d(1/\varepsilon)^d.$$

Taking logarithms one obtains

$$d = \frac{\log N(\varepsilon)}{\log L + \log(1/\varepsilon)},$$

and in the limit of small ε, the term involving L becomes negligible.

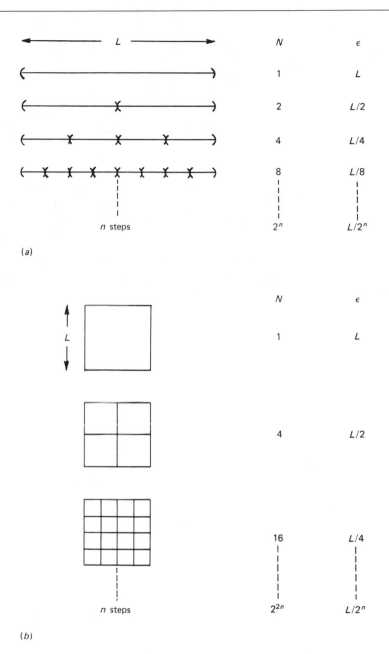

Fig. 5.2 Application of the box covering method to calculate capacity dimension. Boxes of decreasing size ε are used, leading to increasing numbers of boxes. The scaling exponent gives the dimension.

The *capacity* dimension is defined as

$$d_C = \lim_{\varepsilon \to 0} \frac{\log N(\varepsilon)}{\log(1/\varepsilon)}.$$

An equivalent approach is to regard d_C as the slope of the $\log N$ versus $\log(1/\varepsilon)$ curve as $\varepsilon \to 0$. As shown in Figure 5.3 the Cantor set has capacity dimension $d_C = \log 2/\log 3$.

For a dissipative dynamical system such as the pendulum, the attractor resides in an n-dimensional phase space, but its dimension is less than n. For example, the periodic limit cycle of a lightly driven pendulum in (θ, ω, ϕ) space is an elliptical spiral with $d = 1$, as seen in Figure 3.3(a). The chaotic attractor of Figure 3.3(d), consisting of an infinite number of closely spaced sheets, has zero volume and d_C between 2 and 3. Chaotic attractors generally have noninteger dimension, and are called *strange attractors*.

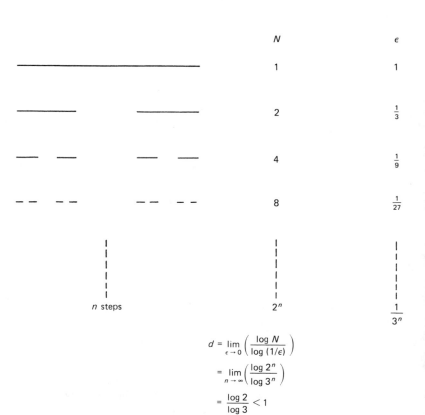

Fig. 5.3 Calculation of the capacity dimension for the Cantor set.

Dimension calculations for dynamical systems with periodic forcing may utilize the Poincaré section rather than the full attractor (Moon and Li, 1985). The time or ϕ direction contributes 1 to the full attractor dimension, and therefore the dimension of the Poincaré section is $D_C = d_C - 1$. Hence, the Poincaré section of the lightly driven, periodic pendulum has $D_C = 0$ (it's just a point), whereas the dimension of the Poincaré section for chaotic states is between 1 and 2. Furthermore, a transverse cut across the closely spaced lines of the Poincaré section reveals a Cantor-like structure. One can think of this transverse direction as contributing the fractional part to the overall fractional dimension of the Poincaré section.

The capacity dimension is only one of several types of dimension (Farmer, Ott, and Yorke, 1983). However, any reasonable definition is expected to have the following properties:

(a) If $A \subset B$ then $d(A) \leq d(B)$. In the case of the pendulum, the Poincaré section is a subset of the full attractor, and, in agreement with this property, the dimension of the section is less (by 1) than that of the attractor.

(b) The null set has zero dimension.

(c) $d(A \times B) = d(A) + d(B)$. (The symbol $\times$ indicates the Cartesian product. If $a = \{x_i \,|\, i = 1, \cdots, n\}$ and $B = \{y_j \,|\, j = 1, \cdots, m\}$ then $A \times B = \{(x_i, y_j) \,|\, i = 1, \cdots, n; j = 1, \cdots m\}$.) For the Cantor set in two-space, the dimension is just the sum of the dimensions of two Cantor sets in one-space:

$$\frac{\log 2}{\log 3} + \frac{\log 2}{\log 3} = \frac{\log 4}{\log 3} \qquad \text{(Problem 3)}$$

(d) If f is a differentiable and invertible function on the set A, then $d(f(A)) = d(A)$. For the pendulum the Poincaré section represents the state at a specific phase in the drive cycle. At a slightly later phase, the section is modified according to some unknown function, f, acting upon the previous state. Numerical simulation shows that the dimension remains constant, in conformity with this property.

The capacity dimension may be estimated from experimental or numerically generated data. (Implicit in such estimates is the assumption that the finite set of data points is only a sampling of an infinite set. If the data were actually limited to a finite set, then its dimension would be zero.) The technique of *box counting*, previously applied to the Cantor set, can be used for experimental or numerical data. In Figure 5.4, a portion of the pendulum's Poincaré section is shown covered with square boxes of length ε. The required number of boxes of size ε is $N(\varepsilon)$, and ε is varied to obtain $N(\varepsilon)$. From the definition given above, d_C is the slope of the graph of $\log[N(\varepsilon)]$ versus $\log[1/\varepsilon]$ as $\varepsilon \to 0$. For this choice of parameters of the pendulum $d_C = 1.3 \pm 0.1$.

For experimental data or higher-dimensional dynamical systems another type of dimension is more efficient to compute than the capacity dimension. This is the *correlation dimension*, d_G (Grassberger and Procaccia, 1983). Suppose that many points are scattered over a set. The typical number of *neighbors* of a given point will vary more rapidly with distance from that point if the set has high dimension than otherwise. The correlation dimension may be computed from the correlation function $C(R)$ defined by

$$C(R) = \lim_{N \to \infty} \left[\frac{1}{N^2} \sum_{i,j=1}^{N} H(R - |\mathbf{x}_i - \mathbf{x}_j|) \right],$$

where $\mathbf{x}_i$ and $\mathbf{x}_j$ are points on the attractor, $H(y)$ is the Heaviside function (1 if $y \geq 0$ and 0 if $y < 0$), and N is the number of points randomly chosen from the entire data set. The Heaviside function simply counts the number of points within a radius R of the point denoted by $\mathbf{x}_i$, and $C(R)$ gives the *average* fraction of points within R.

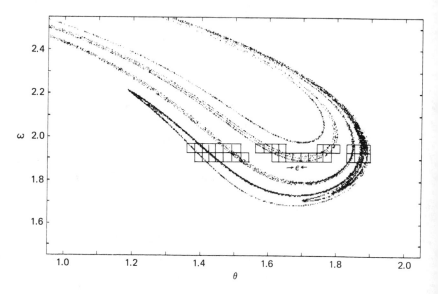

Fig. 5.4 A portion of the pendulum's Poincaré section with some representative boxes covering the points of the section. $g = 1.4954$, $q = 4$.

Figure 5.5 illustrates the method; each point has a circle of radius R drawn around it, and then all points within all circles of size R contribute to $C(R)$. This method of estimating the dimension of a set has the advantage of using less computer memory and computational time. The correlation dimension is defined by the variation of $C(R)$ with R:

$$C(R) \sim R^{d_G} \text{ as } R \to 0.$$

Therefore the correlation dimension is the slope of a graph of $\log C(R)$ versus $\log R$. Figure 5.6 shows such a graph for the driven pendulum, and its straight line portion has a slope $d_G = 1.3 \pm 0.1$. The line differs from a straight line for high and low values of R, for the following reasons. When R approaches the size of the phase space, $C(R)$ saturates at unity since all points are then included in the circle. On the other hand, when R is smaller than the spacing between the data points, only one point lies in each circle, and $C(R)$ levels off at $1/N^2$.

The quantities, d_C and d_G, are not equivalent. The capacity dimension depends only on whether small elements of phase space contain *any* points and does not take into account the differing numbers of points in the various elements. That is, small scale variations of the *density* of points are ignored. On the other hand, the correlation dimension does include this effect. Because of these

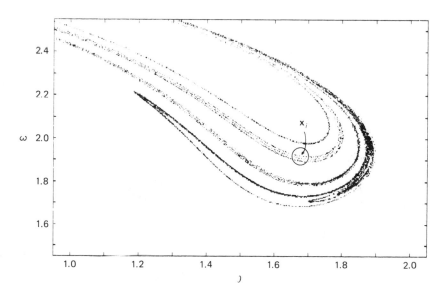

Fig. 5.5 A portion of the Poincaré section with a representative circle centered on the *ith* point. All the other points within the circle are counted by the Heaviside function. $g = 1.4954$, $q = 4$.

differences, d_C is called a *metric* dimension and d_G is called a *frequency dimension*. (See, for example, Farmer *et al.* (1983).)

A third quantity, the *information dimension d_I*, is related to the entropy defined in Chapter 4. Like d_G, the information dimension depends on the distribution of points on the attractor. Suppose the attractor is covered by a set of n boxes of size ε, and let the probability that a point is in the ith box be p_i. Then the metric entropy or missing information (see Chapter 4) is

$$I(\varepsilon) = -\sum_{i=1}^{n} p_i \log p_i.$$

If the points are equally distributed on the fractal, $I(\varepsilon)$ takes its maximum value of $I_0 = \log n$. The information dimension d_I is defined by the equation:

$$d_I = \lim_{\varepsilon \to 0} \left[-\frac{I(\varepsilon)}{\log \varepsilon} \right] = \lim_{\varepsilon \to 0} \left(\frac{\sum_{i=1}^{n} p_i \log p_i}{\log \varepsilon} \right).$$

The information dimension is the 'scaling exponent' in the variation of the entropy, or missing information with ε. The remarkable feature of

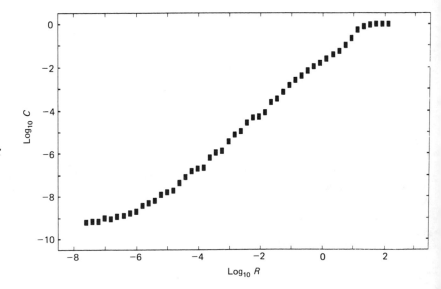

Fig. 5.6 Graph of $\log_{10} C$ versus $\log_{10} R$ for a Poincaré section of the driven pendulum. The correlation dimension is the slope of the straight line portion.

these different definitions of dimension is that they have been shown to be related:

$$d_C \leq d_I \leq d_G.$$

If the points are uniformly distributed on the fractal, the equality holds. In practice the numerical values are usually close together (Grassberger and Procaccia, 1983).

The above definitions of dimension have been consolidated into a general definition of dimensions of *order q*:

$$d^{(q)} = \frac{1}{q-1} \lim_{\varepsilon \to 0} \left(\frac{\log \sum_{i=1}^{n} p_i^q}{\log \varepsilon} \right)$$

where n is the number of phase space elements and p_i is the probability that an attractor point falls in the ith element. (A concise review is given in Atmanspacher, Scheingraber, and Voges (1988).) It can be shown that d_C, d_I, and d_G are equivalent to $d^{(0)}$, $d^{(1)}$, and $d^{(2)}$, respectively (see Problem 9). The higher order dimensions are sensitive to nonuniformities in the distributions of points on the attractor. If they are fairly evenly distributed, then the dimension is unchanged for higher q. In general, $d^{(q)} \geq d^{(q')}$ where $q \leq q'$.

Are so many dimensions really needed to describe an attractor? The answer is yes; strange attractors, like many other fractals, contain more geometrical information than can be described by a single number, in much the same way that the mass distribution of an extended object requires an infinite number of parameters (the 'moments' of a mass distribution) for a complete characterization.

In recent studies of strange attractors, an alternative to the use of these generalized dimensions has emerged. It is based on the idea that complex sets may often be described as 'multifractals', that is, sets consisting of many interwoven fractals, each with a different fractal dimension. The multifractal concept, which is now well substantiated experimentally, leads to characterization of the set by a function called the multifractal spectrum. This function contains the same information as the generalized dimensions $d^{(q)}$, so we will not discuss it in this book. However, an accessible account is provided in Glazier and Libchaber (1988). (See also Halsey *et al.* (1986).)

While the calculation of dimension is a common feature of the research literature on strange attractors, misleading results are often obtained as shown by Arneodo *et al* (1987) and microcomputers are not well suited to these calculations. Therefore we approach the dimension calculation through a remarkable connection between the dimension of the attractor and the time variation of the motion, as characterized by the Lyapunov exponents. In the next two sections we develop this aspect of the subject.

Lyapunov exponents

In Chapter 4 we introduced the idea of the Lyapunov exponent and gave an example for the one-dimensional map. Constraints on the exponents were also given for chaotic, dissipative systems in higher-dimensional spaces. For these systems, of which the pendulum is an example, the conditions are that the sum of all the exponents must be negative, but at least one must be positive to allow sensitive dependence on initial conditions.

For higher-dimensional systems, the calculation of Lyapunov exponents is more challenging than in the one-dimensional case.

Fig. 5.7 A two-dimensional example of the calculation of Lyapunov exponents. (a) The evolution of a sphere of initial points to an ellipsoid. (b) Readjustment of the size of vectors along the principal axes. Beginning at t_1, an orthonormal set of vectors from the center of the sphere evolves by stretching and contracting along the axes of the developing ellipsoid. At t_2, a new set of orthonormal vectors is generated such that one of the new vectors is parallel to the previous stretching direction. The process is repeated over many drive cycles to determine average rates of divergence and convergence of nearby trajectories.

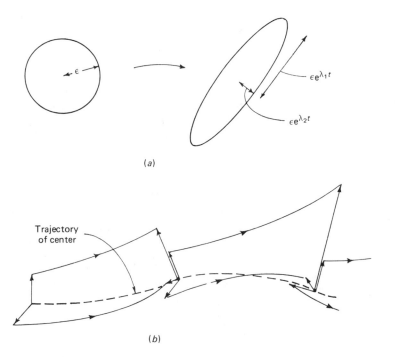

However, the idea is the same: measurement of the average rate of divergence of neighboring trajectories on the attractor. (See Wolf (1986) for a review article which gives an overview of the subject.)

The direction of maximum divergence or convergence is a changing local property on the attractor. The motion must be monitored at each point along the trajectory. Therefore, a small sphere is defined whose center is a given point on the attractor and whose surface consists of phase points from nearby trajectories. As the center of the sphere and its surface points evolve in time, the sphere becomes an ellipsoid, with principal axes in the directions of contraction and expansion. The evolution of the ellipsoid is illustrated in Figure 5.7(a). The average rates of expansion or contraction along the principal axes are the Lyapunov exponents. For the ith principal axis, the corresponding exponent is defined as

$$\lambda_i = \lim_{t \to \infty} \{(1/t)\log[L_i(t)/L_i(0)]\},$$

where $L_i(t)$ is the radius of the ellipsoid along the ith principal axis at time t (Wolf *et al.*, 1985). In this expression, the growth rate is always measured along the ith principal axis, but the absolute orientation in phase space of that axis is not fixed. It is impractical to perform the actual computation in the way suggested by the definition, because the initially close phase points would soon diverge from each other by distances approaching the size of the attractor, and the computation would then fail to capture the local rates of divergence and contraction. Therefore, vectors connecting the surface of the ellipsoid to the center must be shrunk periodically or *renormalized* to ensure that the size of the ellipsoid remains small and that its surface points correspond to trajectories near that of the center point. The renormalization is illustrated in Figure 5.7(b). The Lyapunov exponents are taken to be the *averages* of those obtained over many segments of the central trajectory.

The algorithm for the Lyapunov exponent calculation in more than one dimension is somewhat beyond the scope of this text, although a computer program that calculates the exponents for the pendulum is included in Appendix B. The method is described in some detail in Wolf *et al.* (1985), and FORTRAN code given in that article is the basis for the program listing.

Nevertheless, it is useful to discuss some of the results for the pendulum. Figure 5.8 shows a typical graph of the three computed Lyapunov exponents for the pendulum, plotted as a function of the number of drive cycles utilized for the computation. The initial transients gradually decay as the calculation is extended over many cycles, so that the structure of the attractor is fully explored.

There are three Lyapunov exponents because the pendulum equations have three variables, and their sum should be negative since the system is dissipative. One exponent corresponds to the direction parallel to the trajectory. It contributes nothing to the expansion or contraction of phase volumes, and therefore the corresponding Lyapunov exponent (λ_2) is zero.

The remaining exponents are negative or zero in the periodic states, whereas in the chaotic state one exponent is positive, indicating divergence of trajectories. For the parameters used to generate Figure 5.8 the system is chaotic, and (from the figure) the exponents are estimated to be, $\lambda_1 = 0.16$, $\lambda_2 = 0$, and $\lambda_3 = -0.42$, with an uncertainty of about 0.02.

An interesting relationship can be developed between the Lyapunov exponents and the autonomous set of differential equations describing the pendulum. This relation depends on the fact that both the Lyapunov exponents and the differential equations contain

Fig. 5.8 Graph of the three Lyapunov exponents for the chaotic pendulum versus the number of orbits (or drive cycles) used in the computation. Here $g = 1.4954$ and $q = 4$. After a few cycles, the steady state values are reached.

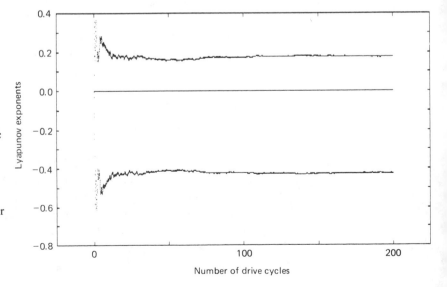

information on the change of volumes in phase space, and the relation provides a check on the numerical values developed from the Lyapunov algorithm. First we note that, according to the definition of Lyapunov exponents, a small volume in phase space will change in time as

$$V(t) = V_0 e^{(\lambda_1 + \lambda_2 + \lambda_3)t},$$

and therefore the rate of change of $V(t)$ is simply

$$dV/dt = \sum_{i=1}^{3} \lambda_i V(t).$$

Furthermore, the discussion of Chapter 2 concerning the change of phase volume in a time δt showed that the time rate of change of the phase volume can also be written as

$$dV/dt = \int \mathbf{\nabla} \cdot \mathbf{F} dV,$$

where $\mathbf{F}$ is the vector function forming the right side of the autonomous set of first order differential equations for the dynamical system. In the case of the pendulum $\mathbf{\nabla} \cdot \mathbf{F}$ equals $-1/q$, independently of position and time. The two expressions for rate of change of volume

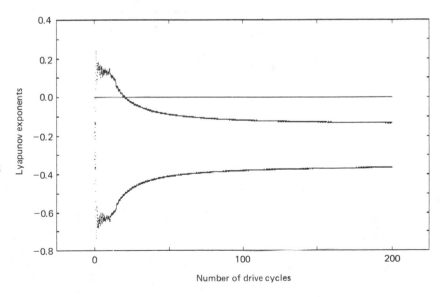

Fig. 5.9 Graphs of Lyapunov exponents computed for a periodic pendulum, with $g = 1.125$ and $q = 2$. The steady state values can be used to check the relation,

$$1/q = -\sum_{i=1}^{3} \lambda_i.$$

are combined and lead to the simple result:

$$1/q = -\sum_{i=1}^{3} \lambda_i.$$

The exponents obtained from Figure 5.8 with $q=4$ satisfy this relation. The Lyapunov exponents for a periodic state (Figure 5.9) also satisfy the relation to within numerical precision.

Lyapunov exponents and dimension

An interesting and important relationship between Lyapunov exponents and dimension was proposed by Kaplan and Yorke (1979). This conjecture can be developed heuristically by reference to Figure 5.10 where a region in (θ,ω) phase space evolves in time under the influence of a dissipative, chaotic system. In this simple example, one direction stretches by a factor of $e^{\lambda_1 t}$ with $\lambda_1 > 0$, and the other direction shrinks according to the factor $e^{\lambda_2 t}$ with $\lambda_2 < 0$. Therefore the area evolves as $A(t) = A_0 e^{(\lambda_1 + \lambda_2)t}$. Now we define a Lyapunov dimension d_L by analogy with the capacity dimension as

$$d_L = \lim_{\varepsilon \to 0} \left[\frac{d(\log N(\varepsilon))}{d(1/\varepsilon)} \right],$$

where $N(\varepsilon)$ is the number of squares with sides of length ε required to cover $A(t)$. Both $N(\varepsilon)$ and the length ε depend on time as follows:

$$N(t) = \frac{A(t)}{\text{square area } (t)} = \frac{A_0 e^{(\lambda_1 + \lambda_2)t}}{A_0 e^{2\lambda_2 t}} = e^{(\lambda_1 - \lambda_2)t}$$

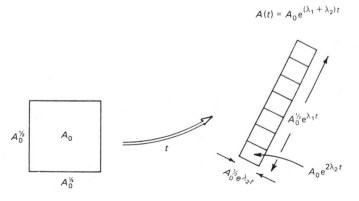

$$A(t) = A_0 e^{(\lambda_1 + \lambda_2)t}$$

Fig. 5.10 Schematic diagram of the role of Lyapunov exponents in the stretching and contraction of an area in phase space.

and

$$\varepsilon(t) = A_0^{\frac{1}{2}} e^{\lambda_2 t}.$$

Combining these expressions with the definition of d_L and using the chain rule, it follows that

$$d_L = 1 - \frac{\lambda_1}{\lambda_2}.$$

This is the Kaplan–Yorke relation. Kaplan and Yorke proposed that $d_L \geq d_G$ and that if the points on the fractal are approximately uniformly distributed, then the equality should hold. (The above derivation depended on a simple example of contraction and expansion in fixed, mutually perpendicular directions. A discussion of the conditions under which the conjecture should be true has been given by Grassberger and Procaccia (1983).)

The Kaplan–Yorke relation may be generalized to higher-dimensional spaces. The formula becomes

$$d_L = j + \frac{\lambda_1 \mid \lambda_2 + \lambda_3 + \cdots + \lambda_j}{|\lambda_{j+1}|},$$

where the λ_i are ordered (λ_1 being the largest) and j is the index of the smallest nonnegative Lyapunov exponent. For example, in the phase space (θ, ω, ϕ) the Lyapunov exponents for the pendulum were found to be $\lambda_1 = 0.16$, $\lambda_2 = 0$, and $\lambda_3 = -0.42$ (when $g = 1.4954$ and $q = 4$). Therefore $j = 2$ and the predicted dimension is

$$d_L = 2 + \frac{0.16 + 0}{0.42} \approx 2.4.$$

In the two-dimensional subspace (θ, ω) the exponents are $\lambda_1 = 0.16$ and $\lambda_2 = -0.42$ and the calculated dimension is $d_L = 1 + \frac{0.16}{0.42} \approx 1.4$, with an uncertainty of 0.05. (Remember that the ϕ dimension corresponds to time and does not contribute to the expansion or contraction of phase volume.) The result $d_L = 1.4$ correlates well with values of both the capacity and correlation dimension calculated directly for this particular state of the pendulum.

The Lyapunov exponents, in conjunction with the Kaplan–Yorke relation, can be used to study chaotic pendulum states with different amounts of damping. For example, when $q = 2$ the pendulum is highly damped, points in the Poincaré section are tightly packed, and the dimension is small. Table 5.1 shows Lyapunov exponents for chaotic

Table 5.1. *Dimensions of the Poincaré section for various damping factors. All states shown are chaotic. Decreasing the damping allows the Poincaré section to spread and the dimension to increase.* $(g = 1.5)$

q	λ_1	λ_2	$\sum_i \lambda_i$	$\dfrac{1}{q}$	$d = 1 + \dfrac{\lambda_1}{\lvert \lambda_2 \rvert}$
4.0	0.16	-0.42	-0.26	0.25	1.38
3.7	0.16	-0.43	-0.27	0.27	1.37
3.0	0.11	-0.44	-0.33	0.33	1.25
2.8	0.09	-0.45	-0.36	0.36	1.2
2.0	0.12	-0.58	-0.46	0.5	1.2

states with various damping factors. The corresponding calculated dimensions are also given. While the changes indicated are not large, there is a general increase in dimension with decreasing damping. However, the pendulum is not chaotic for every value of q in the range shown.

From this discussion it is evident that Lyapunov exponents provide an important link between the fractal geometry of the attractor and the property of sensitive dependence on initial conditions.

Information change and Lyapunov exponents

The condition of sensitivity to initial conditions that is characteristic of chaotic systems implies divergence of initially adjacent dynamical states. While an initial state of the system may be known with a high but finite degree of precision, the ability to predict later states diminishes because of trajectory divergence. Information is lost, or conversely, more information is required to specify the system with the original precision; the entropy has increased.

For many systems the information function has a simple linear time dependence (Atmanspacher and Scheingraber, 1987)

$$I(t) = I_0 + Kt.$$

In this expression $I(t)$ depends on the probabilities p_i, which change with time as the system evolves. Note that I also depends on the length, ε, of the elemental unit that is used to partition the space. The rate of information change or *Kolmogorov entropy* K is an average rate taken over long times T, although in a numerical calculation T would have to be less than the time taken to cover the entire attractor. Subject to these conditions, K is defined as

$$K = \lim_{\varepsilon \to 0} \lim_{T \to \infty} [I(\varepsilon,T)/T].$$

(See, for example, Grassberger and Procaccia (1984).)

As an illustration of Kolmogorov entropy, consider an imaginary dynamical model which is capable of (*a*) deterministic, nonchaotic motion, (*b*) deterministic, chaotic motion, and (*c*) random motion, for which future states are completely unknown. The value of K will depend on the type of motion. We assume that the phase space of the system may be divided into N small regions, each with probability p_i. Suppose that the system is known to be in a particular initial phase region, and that all initial probabilities are zero except the corresponding probability for that region, which is one. Therefore the initial information (entropy) is zero. The three kinds of motion will cause different types of evolution and therefore different changes in the information function.

For non-chaotic time evolution the system's phase trajectories remain close together. After a time T, nearby phase points are closely grouped in some other small region of phase space and there is no change in information. Therefore the Kolmogorov entropy is zero. For chaotic evolution phase trajectories diverge, and the number of phase space regions available to the system after a time T is

$$e^{\lambda_+ T}$$

where λ_+ is a positive Lyapunov exponent. Assuming that all of these regions are equally likely, the information function now becomes

$$I(T) = - \sum_{i=1}^{N} p_i(T) \log p_i(T) = \log e^{\lambda_+ T} = \lambda_+ T.$$

Therefore the Kolmogorov entropy in the chaotic case is λ_+. Finally, for random evolution all phase space regions become possible after a very short time, and assuming equal probabilities, the information entropy is

$I = \log N$

where N is the number of phase space cells. Although N is finite, the fact that each cell has a finite probability of being instantly accessible implies an infinite rate of information expansion. Therefore, in the case of random evolution the Kolmogorov entropy is infinite.

From this discussion it is evident that the Kolmogorov entropy for chaotic systems depends on the positive Lyapunov exponent. In systems with several positive exponents,

$$K \le \sum_{i=1}^{j} \lambda_i$$

where j is the index of the smallest positive λ_i. Grassberger and Procaccia (1983) suggest that equality usually holds. Hence knowledge of the Lyapunov exponents provides a good estimate of Kolmogorov entropy, and positive but finite K implies chaotic behavior. Furthermore, the Kolmogorov entropy provides a means of categorizing the motion of dynamical systems (Atmanspacher and Scheingraber, 1987). If $K = 0$, then the motion is regular (periodic, quasiperiodic, or stationary). If $K > 0$, then the motion is chaotic, and if $K = \infty$, then the motion is random.

Another feature of the Kolmogorov entropy is that it may be used to estimate the time for which future predictions of the state of a chaotic system are valid. After this prediction time, the system's uncertainty has diverged to the size of the phase space (at least in some direction), and the folding action has to occur. Then the system can be described only by a probability density that depends on location in phase space. For purposes of prediction the system becomes probabilistic.

Suppose the system is characterized by a positive Lyapunov exponent, λ_+, and its initial state is defined to within a size ε. Then, in a time, T, the uncertainty in the coordinates will have expanded to the size L of the attractor:

$L \sim \varepsilon e^{\lambda_+ T}$ or $L \sim \varepsilon e^{KT}$.

Either of these relations may be solved for the prediction time:

$T \sim (1/\lambda_+) \log(L/\varepsilon)$ or $T \sim (1/K) \log(L/\varepsilon)$.

Therefore, the prediction time increases only logarithmically with the precision of the initial measurement. For this reason, chaotic states allow only short-term prediction.

Problems

1. The length of the Cantor set may be determined by subtracting the length of each segment taken out of the set during each step in its formation. For example, a length $\frac{1}{3}$ is taken out in the first step, $2(\frac{1}{9})$ in the second step, $4(\frac{1}{27})$ in the third step, and so on. Form the infinite geometric series which this process describes and show that the sum approaches 1. Therefore the Cantor set has no length.

2. Construct a fractal that is similar to the Cantor set, but instead remove the middle $\frac{1}{2}$ from each previous section. Show that its dimension is $\frac{1}{2}$.

3. Construct a 'two-dimensional' Cantor set in the following way. Draw two Cantor sets at right angles to each other so they just touch at one corner. Then fill in those two-dimensional regions where each 'one-dimensional' set intersects the other. In other words form the set which is the Cartesian product of the two original sets, as in Figure (a) below. Show that the capacity dimension is 2(log2/log3). (This set resembles the invariant set of the horseshoe transformation; that is, the points which remain from the original set after many iterations.)

Fig. (a)

4. Following a procedure analogous to that used in Problem 1, calculate the area of the two-dimensional Cantor set defined in Problem 3.

5. Construct a Cantor-like set by taking squares of relative area $\frac{1}{9}$ out of the center of larger squares. The process is illustrated in Figure (*b*). Show that the dimension of the structure is $3(\log2/\log3)$.

6. The Cantor set may be used to study some properties of information dimension. Recall that $I = - \sum_{i=1}^{N} p_i \log p_i$. Calculate I for each state in the development of the set (as shown in Figure 5.1), assuming that each line segment is equally probable at each iteration. Then use the defining expression for information dimension to show that $d_1 = d_C = \log2/\log3$, in this case.

7. Repeat the calculation of Problem 6 but do not assume equal probabilities. At every iteration of the set let the right segment

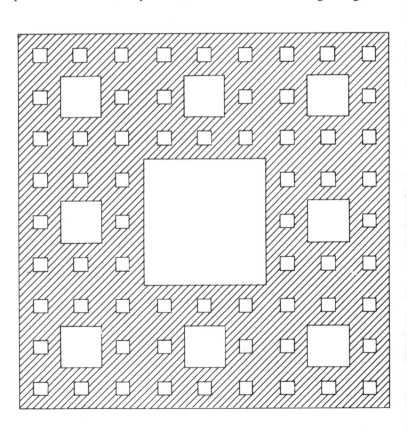

Fig. (*b*)

have twice the probability of the left segment. For example, when there are four segments the probabilities are, from left to right, $\frac{1}{9}, \frac{2}{9}$, $\frac{2}{9}$, and $\frac{4}{9}$. Develop the expression for entropy when there are 2^n segments. Show that, in the limit as n tends to infinity, $d_1 = -1 + 2(\log 2/\log 3)$. (Hint: Be careful with the factorial terms.)

8. By finding the maximum of the function $I = -\sum_{i=1}^{N} p_i \log p_i$, show that d_1 is maximized when $d_1 = d_C$ (See Problem 5 in Chapter 4.)

9. Show that the generalized dimensions are equivalent to d_C and d_1 for $q = 0$ and $q = 1$, respectively. (Hint: For $q = 1$, use L'Hôpital's rule.)

10. For a pendulum with $q = 5$ and $g = 1.5$ the Lyapunov exponents are $\lambda_1 = 0.06$, $\lambda_2 = 0$, $\lambda_3 = -0.26$. Verify the relation between Lyapunov exponents and the damping factor. Calculate the Lyapunov dimension for the attractor in (θ, ω, ϕ) space.

11. Estimate the time for predictability of the pendulum of Problem 10, assuming its state is initially known to within 1% of the range of the phase variables. How would you expect this time to change if the damping were increased?

12. Moon and Li (1985) suggested a way to estimate Lyapunov exponents. The method is based on the fact that at a saddle point the stable and unstable trajectories meet and define directions of convergence and divergence of trajectories, respectively. The rates of these behaviors approximate the respective Lyapunov exponents. To study the behavior of the trajectories in the saddle point region the equation of motion is linearized at the saddle point and then solved. Furthermore it is only necessary to solve the homogeneous part of the differential equation because the saddle point is a fixed point in any Poincaré section taken at frequency ω_D. An examination of Poincaré sections taken at various phases shows that $\theta = \pm\pi$ appears to be the θ saddle point coordinate and therefore the required homogeneous, linearized equation is

$$d^2\theta/dt^2 + (d\theta/dt)/q - \theta = 0.$$

Substitute the solution $\theta = e^{mt}$ into the differential equation to obtain the relation

$$m = -\frac{1}{2q} \pm \left(\frac{1}{4q^2} + 1\right)^{\frac{1}{2}},$$

which has both positive and negative values. These values are the estimates of λ_+ and λ_-. Check that they sum to $\nabla \cdot \mathbf{F}$. Substitute these values into the Kaplan–Yorke relation for d_L, thereby obtaining a relation for d_L in terms of q. Find d_L in the limiting cases where $q = 0$ (infinite damping) and $q = \infty$ (no damping). Calculate d_L in the cases where $q = 2$ and $q = 5$. Do your results match those given in the text and developed in a previous problem? If not suggest a reason. (Moon and Li change the value of the damping parameter to an *effective* value giving numerical results that better match the dimension. If you were to follow a similar procedure, what value of q would you choose for essential agreement with the numerically derived d_L?)

13. Use the program **PENDLYAP** in Appendix B to calculate Lyapunov exponents in various cases. Examine both chaotic and nonchaotic states. Use the bifurcation diagrams to choose the appropriate values of g.

CHAPTER SIX

Concluding remarks

In this book we have discussed many aspects of the dynamics of the driven pendulum. Its behavior is very complex. Variation of parameters leads to an intricate pattern of periodic and chaotic states with several types of transitions between them. In the chaotic regimes, nearby orbits diverge exponentially from each other, leading to unpredictability. Our description has not been exhaustive. For example, fractal basin boundaries were not treated in detail, nor was reference made to an important electronic analogue of the pendulum called the Josephson junction, a superconducting device. Several additional references may be used as starting points in the literature of these subjects (Grebogi, Ott, and Yorke, 1987; Davidson, Dueholm, and Beasley, 1986; Iansiti *et al.*, 1985).

In this chapter we briefly present some examples of chaotic behavior in fluid dynamics, chemical reactions, and laser operation. We then consider the impact of our understanding of chaos on two major fields of physics; quantum physics and statistical mechanics.

Chaos in fluid dynamics

Chaotic motion has been observed prominently in fluids subjected to temperature gradients, differential rotation, vibration, and other forms of energy input. Extensive references may be found in several reviews. (For example, Swinney and Gollub (1986).) A much studied example is Rayleigh–Bénard convection, in which a fluid is placed between two horizontal thermally conducting plates, with the lower one warmer than the upper one, as shown in Figure 6.1. When the

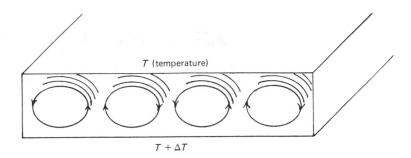

Fig. 6.1 Schematic diagram of Rayleigh–Bénard convection. Arrows indicate the direction of circulation when the temperature difference ΔT is large enough to produce convection.

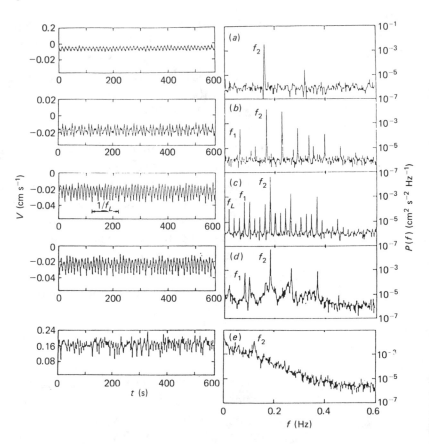

Fig. 6.2 Time series and power spectra of the local fluid velocity showing a sequence (a)–(e) of dynamical states as ΔT is increased. Two distinct oscillations at frequencies f_1 and f_2 develop, then phase lock, and finally lead to chaos in (d). For large ΔT, the spectrum is quite broad. This behavior is analogous to that of a nonlinear circle map (Chapter 4). (From J.P. Gollub and S.V. Benson (1980).)

temperature difference ΔT exceeds a critical value ΔT_c, convection occurs as a series of 'rolls' resembling rotating parallel cylinders. Hot fluid rises, cools, and falls in a spatially periodic pattern. The rolls begin to oscillate transversely in complex ways as ΔT is increased beyond a second threshold $\Delta T_{c2} > \Delta T_c$, and chaotic behavior occurs for even higher values of ΔT. The famous 'Lorenz model' of convection (Lorenz, 1963), though not realistic, was a turning point in the history of nonlinear dynamics. It consists of three coupled ordinary differential equations, and was the first strange attractor to be studied numerically.

Fluid systems are often characterized by experimentally measured time series of the local velocity at a point in the fluid. This can be done remotely using the method known as laser Doppler velocimetry. (For example, see Gollub and Benson (1980).) A laser beam is scattered from the moving fluid: light collected from a chosen point in the cell is slightly Doppler shifted by the moving fluid. Measurement of that small frequency shift (about 1 part in 10^{12}!) gives the instantaneous local fluid velocity. Repetitive measurement yields a time series of the local velocity. Fourier spectra of these time series show the varieties of behavior typically associated with nonlinear systems such as the pendulum, including period-doubling cascades, phase locking between distinct oscillatory modes, and sensitive dependence on initial conditions. One typical route to chaos is shown in Figure 6.2. Two independent oscillations at frequencies f_1 and f_2 develop and interact to produce various sum and difference frequencies. The oscillations phase lock (Figure 6.2(c)) yielding a 'comb' of equally spaced spectral peaks. Chaos appears in (d), and finally a broad nearly featureless spectrum is attained at very large values of ΔT.

Well-defined chaotic states are especially prominent if the horizontal dimensions of the fluid cell are only two or three times its depth. Under these conditions, the fluid behaves as if it has only a few degrees of freedom. It is possible to measure phase space trajectories for this system, and then to determine the dimension of the resulting strange attractor (Malraison *et al.*, 1983). An example of a Poincaré section near the onset of chaos, showing the folding process typical of strange attractors, is given in Figure 6.3. Dimensions less than 5 are typically found near the onset of chaos. Since a fluid continuum in principle has an infinite number of dynamical variables, the 'condensation' of the dynamics onto only a few of them is quite remarkable.

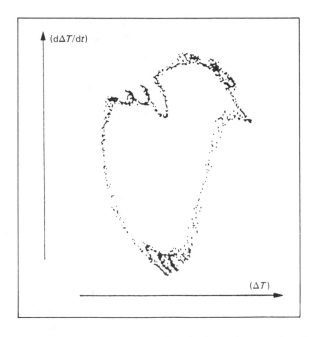

Fig. 6.3 Poincaré section for chaotic thermal convection showing the folding that is characteristic of a strange attractor. The coordinates are the temperature difference ΔT across the fluid layer and its time derivative, both sampled at the frequency of natural oscillation. (From Bergé, Pomeau, and Vidal (1984).) Reprinted by permission of John Wiley and Sons. Copyright 1984 by Hermann, Paris, France.

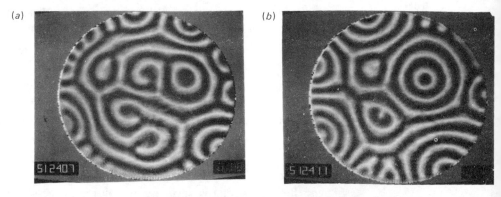

Fig. 6.4 Convective flow patterns in a large circular cell imaged by refraction of a parallel beam of light transmitted through the cell. Dark and light areas correspond to warm (upwelling) and cool (falling) fluid, respectively. The two pictures were made at different times. These flows probably involve many degrees of freedom, and cannot be characterized by a low-dimensional chaotic attractor. (From Heutmaker and Gollub (1987).)

On the other hand, if the horizontal dimensions of the fluid cell are much greater than its depth, the dynamical behavior appears to involve many degrees of freedom. The convective motions then generate complex time-dependent spatial patterns (Figure 6.4) that are certainly unpredictable over long times, but probably cannot be described by orbits in a low-dimensional phase space.

When the temperature difference ΔT is sufficiently large, the motion becomes turbulent. In that case, the motion is nonperiodic at each point in the fluid, and the motions at two different points are uncorrelated with each other. In a chaotic (but nonturbulent) regime, on the other hand, spatial correlations persist. The entire transition process from the quiescent fluid through the chaotic regime to the final turbulent state has been investigated with care (Heslot, Castaing, and Libchaber, 1987) but is still not well understood. In fact, understanding the turbulent state is still one of the major challenges of modern dynamics.

Chaotic chemical reactions

The temporal behavior of chemical reactions is modeled by kinetic rate equations. For the very simple reaction,

$$A \underset{k_r}{\overset{k_f}{\rightleftharpoons}} B,$$

the time dependence of the concentrations is given by

$$\mathrm{d}A/\mathrm{d}t = -k_f A + k_r B$$
$$\mathrm{d}B/\mathrm{d}t = k_f A - k_r B$$

where A and B are used to denote concentrations, k_f is the rate constant for the reaction $A \to B$, and k_r is the rate constant for $B \to A$. For even slightly more complex reactions the equations become complicated and nonlinear. It is therefore expected that under certain conditions these reactions may behave chaotically. For example, a chemical system may exhibit chaotic behavior if kept far from the equilibrium state by a constant infusion of reactants.

The Belousov–Zhabotinskii (BZ) reaction is a much studied example of a chemical system which exhibits both periodic and chaotic behavior. The following simplified model (Swinney, 1983) illustrates some features of the BZ reaction:

$$A + B \underset{k_r}{\overset{k_f}{\rightleftharpoons}} C,$$

The reactants A and B are injected into a closed container with flow rate r, and an exit port relieves the excess material. The rate equations,

$$dA/dt = -k_f AB + k_r C - r(A - A_0)$$
$$dB/dt = -k_f AB + k_r C - r(B - B_0)$$
$$dC/dt = k_f AB - k_r C - rC,$$

exhibit nonlinear coupling between the chemical concentrations. A_0 and B_0 are the reactant concentrations at the input port ($C_0 = 0$). The experimental arrangement is shown in Figure 6.5.

If r is zero the reaction proceeds to equilibrium, and for large r the materials are exhausted from the container before they have time to react. For intermediate r the system has both periodic and chaotic states. In this sense, the input flow rate of reactants is the control parameter analogous to the forcing amplitude of the pendulum. The temperature-dependent rate constants and initial conditions also affect the dynamical state.

A phase space can be constructed for the BZ reaction that allows the periodic and chaotic behaviour to be studied. An example is

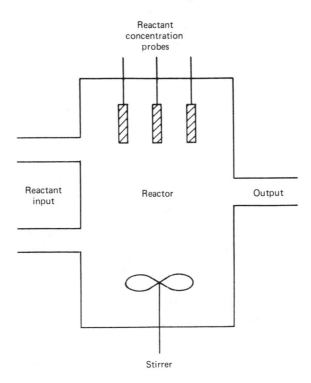

Fig. 6.5 Experimental arrangement of a chemical reaction with reactant flow. The probes monitor the reactant concentrations. (Adapted from Swinney (1983).)

shown in Figure 6.6. The existence of chaos in the BZ reaction suggests that similar behavior might occur for other chemical oscillators, such as those found in biological systems. Chaotic behavior in these systems may indicate a pathological condition, and therefore an analysis of chaotic reactions may prove useful in the study and treatment of disease (Rapp, 1986).

Chaos in lasers

Since the early days of laser technology instabilities in laser action have been apparent. That is, the light output need not be time-independent (Harrison and Biswas, 1986). More recently, efforts have been made to study the chaotic aspects of lasers systematically.

A laser consists of a dielectric material confined between two reflecting mirrors. The energy spectrum of the dielectric must contain two atomic or molecular energy levels whose populations are inverted by an external electrical, optical, or chemical energy source. When population inversion is achieved, photons corresponding to the

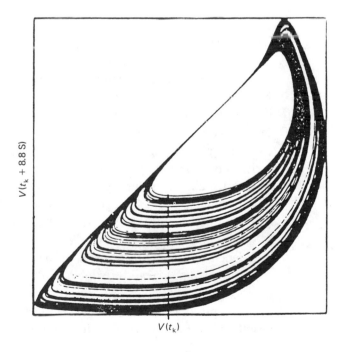

Fig. 6.6 A two-dimensional projection of a three-dimensional phase space construction for the BZ reaction, showing a strange attractor. (From Swinney, 1983). Reprinted by permission of H.L. Swinney.

difference in energy between the two states stimulates atoms in the higher level to decay to the lower level and emit photons. The photons are reflected between the mirrors many times, thus generating the intense electromagnetic field characteristic of the laser.

One model of the laser is a set of differential equations called the Maxwell–Bloch equations. These describe the time dependence of the electric field E, the mean polarization, P, of the atoms, and the amount of population inversion, D. They are:

$$\mathrm{d}E/\mathrm{d}t = -\kappa E + \kappa P$$
$$\mathrm{d}P/\mathrm{d}t = \gamma_1 ED - \gamma_1 P$$
$$\mathrm{d}D/\mathrm{d}t = \gamma_2(\lambda + 1) - \gamma_2 D - \gamma_2 \lambda EP,$$

where κ is the decay rate in the laser cavity due to beam transmission, γ_1 is the decay rate of the atomic polarization, γ_2 is the decay rate of the population inversion, and λ is a pumping energy parameter. (To relate this semiclassical model to a quantum description, note that E is proportional to the expectation value of the photon number density.) The three variables and nonlinear coupling of these equations suggest the possibility of chaotic behavior. Furthermore, the form of these equations is quite similar to the Lorenz model for chaotic convection.

While numerical solutions to the Maxwell–Bloch equations can exhibit chaos, many practical lasers do not operate within a parameter range where chaos occurs. (In many conventional laser configurations, the polarization and population inversion quickly relax to steady values, effectively causing P and D to drop out of the Maxwell–Bloch equations. The equations then do not contain enough variables for chaotic behavior.) However chaotic behavior may be realized when the laser configuration is modified by tuning the cavity length, varying the laser gain, or tilting one of the mirrors. Time series and Fourier spectra of these laser outputs have shown various routes to chaos, including period doubling, intermittency, and quasiperiodicity. (See Gioggia and Abraham (1983) and the review Abraham, Arimondo, and Boyd (1988).)

Special kinds of lasers, such as those where the frequency is broadened by the characteristics of the laser medium (inhomogeneous broadening), readily exhibit both periodic and chaotic behavior (Firth, 1986).

The importance of lasers in modern technology provides a special incentive for the study of their stability. However, the very high frequencies characteristic of laser operation have somewhat impeded the experimental investigation of their dynamical properties.

Chaos and quantum physics

The quantum physics of nonrelativistic systems is based on the Schrödinger equation, a linear differential equation whose solutions give rise to probability distributions for observable quantities. One important feature of quantum physics is the existence of uncertainty relations between certain dynamical variables, such as the position x and momentum p of particles in a collection or ensemble. It is not possible to prepare an initial state in which both of these are well defined. Instead, the collection of particles has distributions of positions and momenta with widths Δx and Δp constrained by Heisenberg's uncertainty relation $\Delta p \Delta x \geq h/4\pi$, where h is Planck's constant. The lack of predictability inherent in quantum mechanics is (at least largely) contained in this mandatory uncertainty in initial conditions.

Quantum systems are not chaotic in the sense used in this book. The Schrödinger equation is linear and yields periodic and quasi-periodic solutions only. Furthermore, the Heisenberg uncertainty relation implies that well-defined trajectories in phase space do not exist over long times. While classical chaos produces the infinite number of infinitely thin layers characteristic of a strange attractor, quantum physics limits the precision of phase trajectories (Gutzwiller, 1985).

On the other hand, quantum systems can sometimes be modeled classically over short time intervals. Bohr's corespondence principle implies that highly excited and closely spaced atomic states near the ionization threshold may be described classically, at least for a limited time during which the probability distributions of the dynamical variables remain localized in phase space. For longer times, the distributions spread and a quantum calculation is required. Even then, the wavefunctions of highly excited states reveal structure that is characteristic of the classical counterpart system. For example, the probability density of a particle in a two-dimensional box shaped like a stadium is enhanced in the neighborhood of unstable periodic orbits of the corresponding classical particle (Heller, 1984).

Semiclassical atomic phenomena may be studied by means of the ionization of electrons from outlying energy states of hydrogen-like atoms (Jensen, 1985). In such states – where the principle quantum number, n, is typically about 60 – the addition of a forcing microwave field causes ionization at a rate that depends strongly on the field amplitude and only weakly on the frequency. This unusual behavior has been successfully simulated by a *classical* model of the outer electron in a one-dimensional Coulomb potential, subject to a time-dependent electric field. In the action-angle phase space of the model, chaotic orbits are observed as the field amplitude is increased through a certain critical value. Trajectories wander over large regions of phase space and provide a diffusion mechanism by which the electrons achieve states of higher energy and eventually ionize. It is quite remarkable that, in fact, the experimental value of the micro-wave field amplitude required for ionization agrees with the threshold value for the onset of chaos in the classical model (Jensen, 1987a).

Foundations of statistical mechanics

The phenomenon of irreversibility as exemplified by the second law of thermodynamics ($\Delta S \geq 0$ for isolated systems) leads to a difficult problem within classical physics. On the one hand, irreversibility implies a preferred direction of time for macroscopic systems. On the other hand, the laws of classical dynamics do not change when the direction of time is reversed. (They are invariant under time reversal.) Therefore the origin of irreversibility in classical dynamics was problematic until Boltzmann proposed a statistical model which accurately predicts macroscopic values of thermodynamic quantities. (A brief history and an elementary version of Boltzmann's original conception is given in Baker (1986).) Boltzmann's model is worth examining here because chaotic dynamics may reduce the need for a statistical assumption.

Boltzmann proposed an explanation of irreversibility for the case of a dilute gas consisting of a large number of hard spheres interacting with each other according to the usual laws of conservation of momentum and energy. The 'gas' is assumed sufficiently dilute that only binary collisions occur. These mechanical aspects of the model seem quite straightforward. The second assumption, perhaps less appealing, is the statistical hypothesis of 'molecular chaos.' After collisions, particles are assumed to lose all memory of their previous velocities. Velocity and position become uncorrelated with each other, and knowledge only of the *distribution* of velocities remains. Since it contains an implicit assumption as to the time direction of events, this statistical hypothesis leads to irreversibility. (A readily accessible demonstration of irreversibility using these assumptions is provided in Baker (1986).)

The discovery of chaotic behavior may render Boltzmann's statistical assumption unnecessary in some cases. Almost 100 years after Boltzmann presented his model, Sinai published the results of an examination of the hard sphere gas as a chaotic system (Sinai, 1970). One can see why this system might be chaotic, as follows. Consider two nearby parallel trajectories for a sphere impinging on one of an array of fixed spheres (see Figure 6.7). The two outgoing trajectories are not parallel, but instead have a small angular divergence $\Delta\theta$. After a second collision the angular divergence is much larger. This leads to exponential growth (on the average) in the angle between the two trajectories, and sensitive dependence on initial conditions. (However there is no dissipation and hence no strange attractor.) The proof that

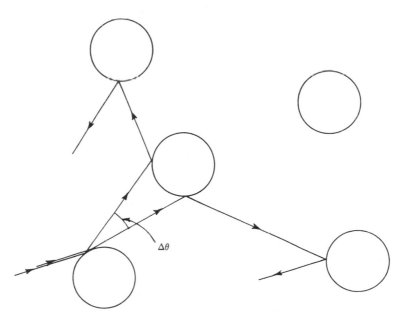

Fig. 6.7 The origin of sensitivity to initial conditions in a hard sphere gas.

the hard sphere gas may be found in all microscopic states $(\mathbf{r}_i, \mathbf{p}_i)$ with roughly equal probability is lengthy and difficult. There is still no proof that many-body systems *in general* will have this property. Still, it is reasonable to think of chaotic dynamics as providing a mechanism for justifying the statistical assumption required for irreversibility.

These brief discussions of the connections of chaos to statistical and quantum physics may be sufficient to show that the foundations of physics are being significantly affected by recent insights in nonlinear dynamics. The impact of chaos on physics may be summarized by the statement that unpredictability enters physics in three major ways: (*a*) through nonlinearity; (*b*) through the uncertainty principle; (*c*) through the statistical behavior of large numbers of particles. However, (*c*) may be in part a consequence of (*a*). The interested reader may pursue the subject further through the references. (For example, see Jensen (1987b).)

The pendulum, one of the paradigms frequently used to represent the Newtonian worldview of the predictable mechanical universe, turns out to be an example of unpredictability. The complexity and richness of its behavior is quite remarkable. Discussed first by Galileo over 400 years ago, the pendulum continues to reveal new aspects of dynamics.

Further reading

The following list of books and papers provides alternate or expanded treatments of many of the subjects in this book. These readings are more accessible to the student and nonspecialist reader than some of the text references to the research literature.

Bak, P., 'The Devil's staircase', *Physics Today*, December 1986, 38–45.

Barcellos, A., 'The fractal geometry of Mandelbrot', *College Mathematics Journal*, **15**, 1984, 98–114. A good introduction that should be read before attempting Mandelbrot's books on fractals.

Bergé, P., Pomeau, Y., and Vidal, C., *Order within chaos*, John Wiley and Sons, Inc., NY, 1984. A comprehensive treatment of chaos, containing a readable account of many aspects of the subject.

Blackburn, J.A., Vik, S., Binruo, Wu, and Smith, H.J.T. 'Driven pendulum for studying chaos', *Rev. Sci. Instrum.* **60**(3), 1989, 422–26. A mechanical model of the pendulum that reproduces the numerical results described in this book.

Chernikov, A.C., Sagdeev, R.Z., and Zaslavsky, G.M. 'Chaos: how regular can it be?', *Physics Today*, November 1988, 27–35.

Crutchfield, P.J., Farmer, J.D., Packard, N.H., and Shaw, R.S., 'Chaos', *Scientific American*, **255**, December, 1986, 46–57.

De Souza-Machado, S., Rollins, R.W., Jacobs, D.T., and Hartman, J.L., 'Studying chaotic systems using microcomputer simulations and Lyapunov exponents', *Am. J. Phys.*, 1990.

Ford, J., 'How random is a coin toss?', *Physics Today*, April, 1983, 40–7.

Glazier, J.A., and Libchaber, A., 'Quasiperiodicity and dynamical systems: an experimentalist's view', *IEEE Trans on Circuits and Systems*, **35**, 1988, 790–809. Though possibly difficult to find in smaller libraries, this paper does have a very readable treatment of generalized dimensions.

Grebogi, C., Ott, E., and Yorke, J.A., 'Chaos, strange attractors, and fractal basin boundaries in nonlinear dynamics', *Science*, **238**, 1987, 632–8.

Hofstader, D.R., 'Metamagical themas', *Scientific American*, **245**, November, 1981, 33–43. A detailed, qualitative discussion of the logistic map and some description of related phenomena.

Holden, V.A., (ed.) *Chaos*, Princeton Univ. Press, Princetown 1986. A collection of review papers on subjects as diverse as Lyapunov exponents and the possibilities of chaos in epidemology.

Jensen, R.V., 'Classical chaos', *American Scientist*, **75**, 1987, 168–81.

Kadanoff, L.P., 'Roads to chaos', *Physics Today*, December 1983, 46–53.

Mandelbrot, B., *The fractal geometry of nature*, W.H. Freeman and Co., NY, 1983.

May, R.M., 'Simple mathematical models with very complicated dynamics,' *Nature*, **261**, 1976, 459–67. The logistic map is treated in great detail, at a somewhat deeper level than that found in Hofstader's paper.

Moon, Francis C., *Chaotic vibrations*, John Wiley and Sons, Inc., NY, 1987. A comprehensive but quite readable treatment of chaos aimed at the applied scientist or engineer. Emphasis is placed on vibrating systems. One of the appendices has some brief, but useful, prescriptions for numerical simulations.

Ottino, J.M., 'The Mixing of fluids', *Scientific American*, **260**, January, 1989, 56–67.

Pool, R., 'Quantum chaos: enigma wrapped in a mystery', *Science*, **243**, February, 1989, 893–5.

Thompson, J.M.T., and Stewart, H.B. *Nonlinear dynamics and chaos*, John Wiley and Sons., Inc., NY, 1986. A comprehensive work aimed at scientists and engineers. The choice of topics is somewhat different from that found in Moon's work, including more emphasis upon stability and bifurcation theory.

Tufillaro, N.B., and Albano, A.M., 'Chaotic dynamics of a bouncing ball', *Am. J. Phys.*, **54**, 1986, 939–44.

Numerical integration – Runge–Kutta method

The mathematical model of the driven pendulum is a nonlinear, second order differential equation, whose solution is obtained numerically. This appendix explains a standard method of numerical integration, the Runge–Kutta method.

As a starting point consider the simple but not very accurate technique called *Euler's method*, and its application to the first order differential equation:

$$dy/dt = f(t,y),$$

with initial conditions for (t,y) of (t_0,y_0). With Euler's method one converts the dt and dy differentials to finite quantities Δt and Δy so that the differential equation becomes

$$\Delta y = \Delta t f(t,y),$$

and this, in turn, may be written as the difference equation:

$$y_{n+1} = y_n + (t_{n+1} - t_n)f(t_n, y_n).$$

Using a fixed interval $\Delta t = (t_{n+1} - t_n)$, the initial values are iterated leading to new (t,y) pairs which eventually form the required solution.

The technique may be extended to higher order differential equations by converting to first order equations. For example suppose the differential equation has the form:

$$d^2y/dt^2 = f(t,y,dy/dt).$$

By the substitution $x = dy/dt$ the following conversion results:

$$dy/dt = x$$
$$dx/dt = f(t,y,x)$$

yielding two first order equations. These can now be replaced by difference

147

equations for the Euler method:

$$x_{n+1} = x_n + (t_{n+1} - t_n)f(t_n, x_n, y_n)$$
$$y_{n+1} = y_n + (t_{n+1} - t_n)x_n.$$

Using the fixed interval, $\Delta t = t_{n+1} - t_n$, the initial values (t_0, x_0, y_0) can be iterated to obtain successive values of (t, x, y).

While conceptually simple, the errors generated by the Euler method grow rapidly with the number of iterations compared to more sophisticated methods. Therefore it should be avoided for extended calculations.

A primary source of error in the Euler technique is that the change in y value is made to depend only on the derivative calculated at the beginning of the Δt step. A better approximation can be found by evaluating the derivative more often during Δt. In the case of a single, first order equation $y' = f(t, y)$ the interval can be split in half and then successive calculations

$$k_1 = \Delta t f(t_n, y_n)$$

and

$$k_2 = \Delta t f(t_n + \Delta t/2, y_n + k_1/2)$$

lead to

$$y_{n+1} = y_n + k_2.$$

This algorithm is the second order Runge–Kutta method, and the process can include more steps for better accuracy. The fourth order Runge–Kutta algorithm is commonly used and seems to represent a reasonable compromise between computer speed and accuracy of solution. The steps in the calculation are as follows:

$$k_1 = \Delta t f(t_n, y_n)$$
$$k_2 = \Delta t f(t_n + \Delta t/2, y_n + k_1/2)$$
$$k_3 = \Delta t f(t_n + \Delta t/2, y_n + k_2/2)$$
$$k_4 = \Delta t f(t_n + \Delta t, y_n + k_3).$$

Then y_{n+1} is evaluated as a weighted average over the k values as:

$$y_{n+1} = y_n + k_1/6 + k_2/3 + k_3/3 + k_4/6.$$

If more than one first order differential equation is to be solved, then the algorithm is applied to each one during each Δt interval. In many of the programs that are listed in Appendix B the subroutine RK4(X,V,TSTEP,XNEW,VNEW,T,W,G,Q), together with the defined function (ACCEL(X,V,T,W,G,Q), does this procedure.

It is important to keep the step interval Δt small enough to maintain accuracy in the solution. Yet if Δt is smaller than necessary the computation will involve excessive time. Furthermore the requisite size of Δt may vary with the rate at which the solution itself varies. These requirements suggest that Δt should be made adjustable during the calculation. One way to do this is to tie the step size to the computed difference Δy. If the Δy is greater than a certain predetermined value ε, then the program can back up and use a smaller Δt to evaluate y_{n+1}. In the programs used in this work Δt was increased with the formula:

$$\Delta t = \Delta t(0.95)(\varepsilon/\Delta y)^{\frac{1}{4}}$$

when

$\Delta y < \varepsilon$, and decreased according to the formula:

$$\Delta t = \Delta t(0.95)(\varepsilon/\Delta y)^{\frac{1}{5}}$$

when $\Delta y > \varepsilon$.

Most of the pendulum programs listed in Appendix B use the Runge–Kutta method with the adaptive stepsize. Other more sophisticated algorithms are available in books on numerical methods, such as Press *et al.* (1986).

Computer program listings

This appendix provides several listings which may be used in their present or modified versions for exercises in the text. The listings are in the language True BASIC™ with the software and hardware requirements described in the Preface. Readers who wish to work in a different language may translate the given code.

The programs are of two types: those which solve the differential equations of the pendulum, and those which iterate discrete maps. Because the Runge–Kutta algorithm is complex, the computer processing time for differential equation solutions is much longer than for the iteration of maps. The reader is warned that some of the pendulum text diagrams took hours and even days to generate with an AT-IBM compatible machine.

The computer exercises of Chapter 2 are based upon the programs, PENDULUM, POINCARE, and EXPFFT. PENDULUM, Listing 1, provides a two-dimensional phase plane representation of the pendulum. The differential equation is that of the non-linear, damped, driven pendulum. Modification of the external function ACCEL(X,V,T,W,G,Q) is required to linearize the pendulum. The term $-X$ would then be substituted for the term $-\mathrm{SIN}(X)$. The subroutine RK4(X,V,TSTEP,XNEW,VNEW,T,W,G,Q) is the Runge–Kutta integrator that is common to all the pendulum programs. Aside from input and graphics statements, the program consists of repeated callings of RK4 and the subsequent plotting of phase points at times which are consistent with the variable step-size procedure outlined in Appendix A. The angular coordinate is kept periodic at each plotting by adjusting, if necessary, its absolute value to a number less than π. Modification of the program to show the evolution of a block of initial θ and ω coordinates requires a loop through the main part of the program. The initial coordinates, denoted XINT and VINT then take successive values for each initial coordinate pair in the block.

The Poincaré section program, POINCARE (Listing 2), is quite similar to

PENDULUM. However, since phase points are not generally calculated at the precise point of section, an additional procedure is required. For each pair of two consecutive phase points a check is made as to whether they straddle (in time) the moment of section. If they do satisfy this condition, the time variable is backed up to the precise moment of section by interpolation. The phase point computation is made at that moment, and the phase point is then plotted as part of the Poincaré section.

The program EXPFFT, Listing 3, computes the power spectrum of a linear combination of periodic components. It will provide both a time series and the power spectrum. A cursor will appear on the screen following the time series display. Any key may be pressed to continue the processing. For the most reliable spectrum, 4096 points of the time series should be used. To obtain the spectrum of some function other than the periodic function, one must change the dozen lines of code following the comment line, '!DEVELOP PERIODIC TIME SERIES DATA'.

The problems of Chapter 3 utilize the programs MOTION (Listing 4), BIFURCATION (Listing 5), FFT (Listing 6), and Basins (Listing 7), as well as PENDULUM and POINCARE. MOTION is an animation of the pendulum motion in 'real' space and time. The program uses RK4 to solve the pendulum equations at equal time intervals. Animation is achieved by rotating the 'pendulum' image through the appropriate angle at each time step. The angular velocity is therefore determined by the computer processing speed.

BIFURCATION generates a 'bifurcation diagram' or scatter plot of angular velocity at closely spaced forcing amplitudes, g. The angular velocity values are taken at fixed phase $\phi = 0$, so the resulting diagram is equivalent to a sequence of Poincaré sections at increasing g values. The processing time is quite long.

FFT calculates the power spectrum of a time series of either $\theta(t)$ or $\omega(t)$. The program first displays a phase diagram of θ versus ω, then a time series of the chosen quantity, and finally, the appropriate power spectrum. The time series algorithm is similar to that used for the Poincaré section in that the dynamical variable must be calculated at precise times, as regulated by the maximum frequency for the power spectrum. At various stages the program requires inputs, either to proceed to the next step, or to specify a variable and type of display. The user may choose the 'Hanning' option which has the ability to diminish spurious sidebands caused by the finite length of the data record. As with EXPFFT, the largest number of points, 4096, is recommended for the most accurate spectrum.

BASINS generates a display of the basins of attraction of the pendulum for a given driving force amplitude. In addition to incorporating the phase plane algorithm of PENDULUM, this program computes the average pendulum velocity $\langle\omega\rangle$, over the time interval ($t$min, $t$max), for each pair of initial coordinates of the phase plane. The sign of $\langle\omega\rangle$ determines the basin toward which the phase trajectory tends from an intial point. If $\langle\omega\rangle$ is positive, a circle is placed at the corresponding initial point. Otherwise the point is unmarked. This program also provides the option to superpose the Poincaré sections on the basins. As with BIFURCATION the processing time of BASINS is quite long.

These programs may be modified to display graphs of other dynamical systems. Specific changes would be required in the ACCEL function, in the parameters of the graphs, and in the input statements.

The program listings for the problems of Chapter 4 are entitled LOGISTIC MAP (Listing 8), CIRCLE MAP (Listing 9), and HENON MAP (Listing 10). LOGISTIC MAP is a multipurpose program that gives the user four display options: a return map, a bifurcation diagram, an entropy diagram, and a Lyapunov exponent diagram. With the possible exception of the entropy algorithm, none of the routines is especially complex. The entropy calculation keeps track of the number of $x(n)$ values that end up in each equally-sized cell of the x interval, $[0,1]$. The entropy is then obtained from the relative frequencies for each cell. LOGISTIC MAP may be readily modified to display the Tent map (Problem 6 of Chapter 4) – primarily by changing the external function which specifies the particular map.

CIRCLE MAP is a multipurpose program that gives the user three options: a return map, a bifurcation map, and a Devil's staircase. As with LOGISTIC MAP, the three calculation subroutines are relatively straightforward. HENON MAP provides an illustration of a two-dimensional map.

The final listing, PENDLYAP (Listing 11), calculates the Lyapunov exponents of the driven pendulum, as discussed in Chapter 5. The code is an adaptation of a FORTRAN program for the exponents of the Lorenz system of equations. (See the appendix of Wolf *et al.* (1985).) The resultant graph illustrates the gradual convergence of the exponents to their respective values.

All of the above programs and several others are available on diskette as a menu-driven runtime package for IBM compatible PCs. This diskette obviates the need for any True BASIC software or typing of the listings. The order form at the back of the book may be used to purchase the package.

Listing I

```
!PROGRAM TITLE - *******PENDULUM********
!THIS PROGRAM DISPLAYS THE 2-DIMENSIONAL PHASE DIAGRAM
!FOR THE DRIVEN AND UNDRIVEN PENDULUM.
!
LIBRARY "SGLIB.TRC"
DECLARE DEF ACCEL
DIM A(1), B(1)
!INPUT STATEMENTS
 INPUT PROMPT"INPUT DRIVING FORCE STRENGTH:":g
 INPUT PROMPT"INPUT DAMPING (IF NO DAMPING THEN INPUT 9999999):":q
 INPUT PROMPT"INPUT: INITIAL ANGLE , ANGULAR VELOCITY:":XINT,VINT
 INPUT PROMPT"INPUT: MINUMUM TIME , MAXIMUM TIME:":TMIN,TMAX
 INPUT PROMPT"AVERAGE VELOCTY CALCULATION; YES(1) , NO(2):":AVC
CALL PARAMS(W,EPS,TSTEP,XMIN,XMAX,YMIN,YMAX)   !SETS MISC AND GRAPH PARAMETERS
CALL SETXSCALE(XMIN,XMAX)   !FROM SGLIB
CALL SETYSCALE(YMIN,YMAX)   !FROM SGLIB
CALL SETTEXT("PENDULUM - 2-D PHASE DIAGRAM","ANGLE","ANGULAR VELOCITY")
CALL RESERVELEGEND  !FROM SGLIB , SAVES SPACE FOR LEGENDS

DATA 0,0
CALL DATAGRAPH(A,B,1,0,"WHITE")  !FROM SGLIB - PLOTS INITIAL POINT
LET T=0
LET X=XINT
LET V=VINT
CALL GOTOCANVAS   !SETS SCREEN FOR GRAPH
!
!CALCULATION AND GRAPHNG BLOCK
FOR I=1 TO 10000000
   CALL RK4(X,V,TSTEP,XNEW,VNEW,T,W,G,Q)   !CALL RUNGE-KUTTA, STEP = TSTEP
    LET TSHALF=TSTEP/2    ! SPLIT INTERVAL
   CALL RK4(X,V,TSHALF,XNH,VNH,T,W,G,Q)   ! DO TWO HALF STEPS
   CALL RK4(XNH,VNH,TSHALF,XN,VN,T+TSHALF,W,G,Q)
   LET D1=ABS(XN-XNEW)
   LET D2=ABS(VN-VNEW)
   LET DELTA=MAX(D1,D2)
   IF DELTA<EPS THEN
     IF T>TMIN THEN
       IF ABS(X)>PI THEN LET X=X-2*PI*ABS(X)/X
       CALL GRAPHPOINT(X,V,1)
       LET SUMVEL=SUMVEL+V*TSTEP    !UPDATE AVERAGE
     END IF
     LET X=XNEW
     LET V=VNEW
     LET T=T+TSTEP
     LET TSTEP=TSTEP*.95*(EPS/DELTA)^.25
     IF ABS(X)>PI THEN LET X=X-2*PI*ABS(X)/X
   ELSE
     LET TSTEP=TSTEP*.95*(EPS/DELTA)^.2   !REDUCE STEP SIZE
   END IF
   IF T>TMAX THEN LET I=10000001
   NEXT I
   LET MEANVEL=SUMVEL/(TMAX-TMIN)
   CALL ADDLEGEND("G="&STR$(G)&"    Q="&STR$(Q),0,1,"WHITE")
   IF AVC=1 THEN CALL ADDLEGEND("AV. VEL. = "&STR$(MEANVEL),0,1,"WHITE")
   CALL DRAWLEGEND     !ADDS G AND Q VALUES TO LEGEND
get key variable
clear
print"press <esc> key to finish"
END
!
SUB RK4(X,V,TSTEP,XNEW,VNEW,T,W,G,Q)     !RUNGE-KUTTA INTEGRATOR
   DECLARE DEF ACCEL
   LET XK1=TSTEP*V
   LET VK1=TSTEP*ACCEL(X,V,T,W,G,Q)
   LET XK2=TSTEP*(V+VK1/2)
   LET VK2=TSTEP*ACCEL(X+XK1/2,V+VK1/2,T+TSTEP/2,W,G,Q)
   LET XK3=TSTEP*(V+VK2/2)
   LET VK3=TSTEP*ACCEL(X+XK2/2,V+VK2/2,T+TSTEP/2,W,G,Q)
```

```
      LET XK4=TSTEP*(V+VK3)
      LET VK4=TSTEP*ACCEL(X+XK3,V+VK3,T+TSTEP,W,G,Q)
      LET VNEW=V+(VK1+2*VK2+2*VK3+VK4)/6
      LET XNEW=X+(XK1+2*XK2+2*XK3+XK4)/6
END SUB
!
DEF ACCEL(X,V,T,W,G,Q)
      LET DAMP=1/Q
      LET ACCEL=-SIN(X)-DAMP*V+G*COS(W*T)
END DEF
!
SUB PARAMS(W,EPS,TSTEP,XMIN,XMAX,YMIN,YMAX)
  LET W=0.66666666
  LET EPS=1.0E-6
  LET TSTEP=0.5
  LET XMIN=-3
  LET XMAX=3
  LET YMIN=-3
  LET YMAX=3
END SUB
```

Listing 2

```
!PROGRAM TITLE **************POINCARE**************
CLEAR
PRINT"                 ***PENDULUM - POINCARE SECTION***"
PRINT
PRINT"THIS PROGRAM DISPLAYS THE POINCARE SECTION OF THE PENDULUM"
PRINT"AND CAN SAVE THE DATA TO A FILE."
LIBRARY "SGLIB.TRC"
!
DECLARE DEF accel
DIM A(1),B(1)
INPUT prompt"Input driving force strength: ":g
INPUT prompt"Input damping (If no damping then input 9999999):":q
INPUT prompt"Input initial angle, angular velocity: ": xint,vint
INPUT Prompt"Input min. and max. time:":tmin,tmax
INPUT prompt"Input phase angle/(2*pi): ":phi
INPUT PROMPT" SAVE DATA TO A FILE? YES(1), NO(2):":SAVEFILE
IF SAVEFILE=1 THEN
  INPUT PROMPT"FILE NAME FORMAT EX. 14954020 :":FILENAME
  INPUT PROMPT"DRIVE FOR FILE DISK A,B,C,ETC.:":DISK$
  LET NAME$=DISK$&":"&STR$(FILENAME)
END IF
!
CALL PARAMS(W,EPS,TSTEP,XMIN,XMAX,YMIN,YMAX)
CALL SETXSCALE(XMIN,XMAX)
CALL SETYSCALE(YMIN,YMAX)
CALL SETTEXT("PENDULUM POINCARE SECTION","THETA","OMEGA")
CALL RESERVELEGEND
!
DATA 0,0
CALL DATAGRAPH(A,B,1,0,"RED")
LET t=0
LET x=xint
LET v=vint
CALL GOTOCANVAS
!
!CALCULATION AND GRAPHING BLOCK
LET phi=phi*2*pi
IF SAVEFILE=1 THEN
OPEN #1:NAME NAME$, ORGANIZATION RECORD, CREATE NEWOLD
ASK #1:FILESIZE LENGTH
IF LENGTH=0 THEN SET#1:RECSIZE 10
SET #1: POINTER END
END IF
FOR i=1 to 1000000
```

```
      CALL rk4(x,v,tstep,xnew,vnew,t,w,g,q)         ! Take a step of size tstep
      LET tshalf=tstep/2
      CALL rk4(x,v,tshalf,xnh,vnh,t,w,g,q)          !Take two half steps

      CALL rk4(xnh,vnh,tshalf,xn,vn,t+tshalf,w,g,q)
      LET d1=abs(xn-xnew)
      LET d2=abs(vn-vnew)
      LET delta=max(d1,d2)
      IF delta<eps then
         IF t>tmin then
            LET tnew=t+tstep
            LET w1=mod(phi-w*t,2*pi)        !Check for Poincare section
            LET w2=mod(w*tnew-phi,2*pi)
            IF w1<w*tstep then
               IF w2<w*tstep then
                  LET ts=w1/w
                  CALL rk4(x,v,ts,xp,vp,t,w,g,q)   !CALCULATES POINT AT SECTION
                  IF abs(xp)>pi then LET xp=xp-2*pi*abs(xp)/xp
                  CALL GRAPHPOINT(XP,VP,1)
                  IF SAVEFILE=1 THEN WRITE #1:XP,VP
               END IF
            END IF
         END IF
         LET x=xnew
         LET v=vnew
         LET t=t+tstep                  !Expand step size
         LET tstep=tstep*.95*(eps/delta)^.25
         IF abs(x)>pi then              !bring theta back into range
            LET x=x-2*pi*abs(x)/x
         END IF
      ELSE                              !else reduce step size
         LET tstep=tstep*.95*(eps/delta)^.2
      END IF
      IF t>tmax then LET i=1000001
NEXT i
LET G$=STR$(G)
LET Q$=STR$(Q)
CALL ADDLEGEND("G="&STR$(G)&"     Q="&STR$(Q)&"     PHI="&STR$(PHI),0,1,"WHITE")
CALL DRAWLEGEND
END
!
!
!
SUB rk4(x,v,tstep,xnew,vnew,t,w,g,q)
   DECLARE DEF accel
   LET xk1=tstep*v
   LET vk1=tstep*accel(x,v,t,w,g,q)
   LET xk2=tstep*(v+vk1/2)
   LET vk2=tstep*accel(x+xk1/2,v+vk1/2,t+tstep/2,w,g,q)
   LET xk3=tstep*(v+vk2/2)
   LET vk3=tstep*accel(x+xk2/2,v+vk2/2,t+tstep/2,w,g,q)
   LET xk4=tstep*(v+vk3)
   LET vk4=tstep*accel(x+xk3,v+vk3,t+tstep,w,g,q)
   LET vnew=v+(vk1+2*vk2+2*vk3+vk4)/6
   LET xnew=x+(xk1+2*xk2+2*xk3+xk4)/6
END SUB
DEF accel(x,v,t,w,g,q)
   LET accel=-sin(x)-(1/q)*v+g*cos(w*t)
END def
!
SUB PARAMS(W,EPS,TSTEP,XMIN,XMAX,YMIN,YMAX)
   LET W=0.66666666
   LET EPS=1.0E-6
   LET TSTEP=0.5
   LET XMIN=-3
   LET XMAX=3
   LET YMIN=-3
   LET YMAX=3
END SUB
```

Listing 3

```
1000 !PROGRAM TITLE ********EXPFFT*******
1010 LIBRARY "SGLIB.TRC"
1011 CLEAR
1012 PRINT"                  ***FFT OF SUPERPOSED SINE WAVES***"
1013 PRINT
1014 PRINT"THIS PROGRAM TAKES THE FOURIER TRANSFORM OF A GROUP OF SINE"
1015 PRINT"WAVES WHOSE AMPLITUDES AND FREQUENCIES ARE INPUTS.  BOTH THE"
1016 PRINT"TIME SERIES AND THE TRANSFORM ARE GRAPHED.  IF A BLINKING CURSOR"
1017 PRINT"APPEARS PRESS ANY KEY TO CONTINUE.  THE HANNING OPTION SMOOTHS "
1018 PRINT" THE ABRUPT EFFECT OF THE WINDOW "
1019 PRINT" AND SUPPRESSES SPURIOUS COMPONENTS. "
1030 DIM thetadata(5000),thetadotdata(5000),xreal(0 to 5000),ximag(0 to 10000)
1040 DIM tpoint(0 to 5000),power(2048),frequency(2048),FREQ(10),AMPL(10)
1050 DECLARE DEF bitr
1060 INPUT prompt"Max frequency : ":maxfreq
1070 INPUT Prompt"Input min.time:":tmin
1080 INPUT prompt"No. of FFT points(..256,512,1024,2048,4096) : ":number
1090 LET ps=1
1100 LET del=.5/maxfreq
1110 LET tmax=number*del+tmin
1120 LET n=number
1130
1140 LET count=0
1150 LET p=1
1160 !DEVELOP PERIODIC TIME SERIES DATA
1170 INPUT PROMPT "HOW MANY FREQUENCY COMPONENTS?":NUMBFREQ
1180 FOR NF = 1 TO NUMBFREQ
1190     INPUT PROMPT" STATE FREQUENCY :":FREQ(NF)
1200     INPUT PROMPT" COMPONENT AMPLITUDE (LESS THAN 1):":AMPL(NF)
1210 NEXT NF
1220 FOR P = 1 TO N
1230     LET TOTAL = 0
1240     FOR NF = 1 TO NUMBFREQ
1250         LET TOTAL = TOTAL + AMPL(NF)*SIN(2*PI*FREQ(NF)*(TMIN+P*DEL))
1260     NEXT NF
1270     LET THETADOTDATA(P)=TOTAL
1280 NEXT P
1290 !
1300 !
1310 !PREPARATION OF THE FFT DATA
1320 CLEAR
1330 INPUT prompt" HANNING OPTION Y/N? ": hanning$
1340 LET tgamma=log2(n)
1350 IF abs(int(tgamma)-tgamma)=0 then
1360     LET gamma=tgamma
1370     GOTO 1400
1380 END IF
1390 LET gamma=int(tgamma)+1
1400 PRINT "gamma= ";gamma
1410 LET newn=2^gamma
1420 LET nu=gamma
1430 FOR i=n+1 to newn
1440     LET xreal(i)=0
1450 NEXT i
1460 LET n=newn
1470 PRINT"n=";n
1480 CLEAR
1490 IF ps=1 then LET title$="WAVE DISPLACEMENT"
1500 CALL settext("TIME SERIES","TIME",title$)
1510 CALL setxscale(tmin,tmax)
1520 FOR k=0 to n-1
1530
1540     IF ps = 1 then
1550         LET xreal(k)=thetadotdata(k+1)
1560     ELSE IF ps = 2 then
1570         LET xreal(k)=thetadata(k+1)
1580     END IF
1590     IF hanning$="y" then LET xreal(k) = xreal(k)*(.5-.5*cos(2*pi*k/(n-1)))
```

```
1600      LET ximag(k)=0
1610      LET tpoint(k)=tmin+k*del
1620 NEXT k
1630 CALL SETAXES(0)
1640 CALL setgraphtype("")
1650 CALL datagraph(tpoint,xreal,1,0,'white')
1660 GET KEY keyvariable
1670 FOR i= 1 to 100
1680 NEXT i
1690
1700 !FFT ALGORITHM
1710 CLEAR
1720 PRINT "Calculating FFT"
1730 LET n2=n/2
1740 LET nu1=nu-1
1750 LET k=0
1760 FOR l=1 to nu
1770      DO while k<(n-1)
1780         FOR i=1 to n2
1790            LET argument=k/2^nu1
1800            LET garbage=int(argument)
1810            LET p=bitr(garbage,nu)
1820            LET arg =2*pi*p/n
1830            LET c=cos(arg)
1840            LET s=sin(arg)
1850            LET k1=k+1
1860            LET k1n2=k1+n2
1870            LET treal=xreal(k1n2)*c+ximag(k1n2)*s
1880            LET timag=ximag(k1n2)*c-xreal(k1n2)*s
1890            LET xreal(k1n2)=xreal(k1)-treal
1900            LET ximag(k1n2)=ximag(k1)-timag
1910            LET xreal(k1)=xreal(k1)+treal
1920            LET ximag(k1)=ximag(k1)+timag
1930            LET k=k+1
1940         NEXT i
1950         LET k=k+n2
1960      LOOP
1970      LET k=0
1980      LET nu1=nu1-1
1990      LET n2=int(n2/2)
2000 NEXT l
2010
2020 FOR k=1 to n
2030      LET i=bitr(k-1,nu)+1
2040      IF i<=k then GOTO 2110
2050      LET treal=xreal(k)
2060      LET timag=ximag(k)
2070      LET xreal(k)=xreal(i)
2080      LET ximag(k)=ximag(i)
2090      LET xreal(i)=treal
2100      LET ximag(i)=timag
2110 NEXT k
2120
2130 !GRAPHING THE FFT
2140 CLEAR
2150 INPUT prompt"Plot 1)power spectrum, or 2)log power spectrum: ":pps
2160 INPUT prompt"Frequency variable - 1)linear, or 2)log: ":freqvar
2170 LET maxfreq=.5/del
2180 LET minfreq=1/(number*del)
2190
2200 CLEAR
2210 !Y-AXIS
2220 IF pps = 1 then
2230    LET TITLE$="POWER SPECTRUM"
2240    LET YAXIS$="POWER"
2250 ELSE
2260    LET TITLE$="LOG POWER SPECTRUM"
2270    LET YAXIS$="LOG POWER"
2280 END IF
2290 !X-AXIS
2300 IF freqvar=2 then
2310    LET XAXIS$="LOG FREQUENCY"
2320 ELSE
```

```
2330    LET XAXIS$="FREQUENCY"
2340 END IF
2350
2360 !DRAW AXES
2370 CLEAR
2380 CALL setxscale(minfreq,maxfreq)
2390 CALL setyscale(1e-6,.99)
2400 CALL SETTEXT(TITLE$,XAXIS$,YAXIS$)
2410 CALL RESERVELEGEND
2420
2430 !PLOT POINTS
2440 FOR i=1 to n/2
2450    LET frequency(i)=i/(n*del)
2460    LET power(i)=(((xreal(i))^2+(ximag(i))^2))/(n^2)
2480
2490 NEXT i
2500 !PLOT TEXT
2510 CALL setaxes(0)
2520 IF pps=1 then
2530    IF freqvar=1 then CALL setgraphtype("xy")
2540    IF freqvar=2 then CALL setgraphtype("logx")
2550 END IF
2560 IF pps=2 then
2570    IF freqvar=1 then CALL setgraphtype("logy")
2580    IF freqvar=2 then CALL setgraphtype("logxy")
2590 END IF
2595 IF NUMBER =4096 THEN
2596    LET SYMBOL=1
2597 ELSE
2598 LET SYMBOL=0
2599 END IF
2600 CALL datagraph(frequency,power,1,SYMBOL,"white")
2610 CALL ADDLEGEND("N="&STR$(N)&"   MAX FREQ="&STR$(MAXFREQ)&"   DEL F="&STR$(MIN
FREQ),0,1,"WHITE")
2620 CALL drawlegend
2630 IF hanning$="y" then
2640    CALL ADDLEGEND("  HANNING",0,1,"WHITE")
2650 END IF
2660 GET KEY keyvariable
2670 INPUT PROMPT "Another with Hanning? y/n: ":hann$
2680 IF hann$= "y" THEN GOTO 1320
2690 INPUT PROMPT "Different presentation of same FFT? (y/n): ":diffplot$
2700 IF diffplot$ ="y" THEN GOTO 2140
2710 END
2720 !
2730 !BIT REVERSER FUNCTION
2740 DEF bitr(j,nu)
2750    LET j1=j
2760    LET ibitr=0
2770    FOR i=1 to nu
2780        LET j2 = int(j1/2)
2790        LET ibitr=ibitr*2+(j1-2*j2)
2800        LET j1=j2
2810    NEXT i
2820    LET bitr=ibitr
2830 END DEF
```

Listing 4

```
!
!PROGRAM TITLE ***************MOTION***************
!LIBRARY "SGLIB.TRC"
DIM A(1),B(1)
!
CLEAR
print"                    ***PENDULUM - ANIMATION OF ITS MOTION***"
PRINT"This program draws the motion of the pendulum at equal time intervals."
PRINT
```

```
INPUT prompt"Input driving force strength:":g
INPUT prompt"input damping ,q:":q
INPUT prompt"Input initial position:":xint
INPUT Prompt"Input initial velocity:":vint
INPUT prompt"Input min. and max. time:":tmin,tmax
INPUT prompt"Input drive angular frequency:":w
LET tstep=.5

SET WINDOW -1,1,-1,1
BOX LINES -.95,.95,-.95,.95
!CALL SETXSCALE(-1,1)
!CALL SETYSCALE(-1,1)
!CALL SETAXES(0)
!CALL SETTICKSIZES(0,0)
!CALL SETTITLE("PENDULUM ANIMATION")
!DATA 0,0
!CALL DATAGRAPH(A,B,0,0,"WHITE")
!CALL GOTOCANVAS
PICTURE Pendulum
    SET COLOR "white"
    FOR k=1 to 2
        IF k=2 then SET COLOR "black"
        PLOT LINES:0,0;0,-.5;.05,-.5;.05,-.55;-.05,-.55;-.05,-.5;0,-.5
    NEXT k

END PICTURE

FOR i=1 to 1000000
    CALL rk4(x,v,tstep,xnew,vnew,t,w,g,q)
    LET t=t+tstep
    LET x=xnew
    LET v=vnew
    IF t>tmin    then
        LET angle=x
        DRAW pendulum with rotate(angle)
        PLOT
    END IF
    IF t>tmax then LET i=1000001
NEXT i
get key variable
CLEAR
END

DEF accel(x,v,t,w,g,q)
    LET accel= -sin(x)-(1/q)*v+g*cos(w*t)
END DEF

SUB rk4(x,v,tstep,xnew,vnew,t,w,g,q)
    DECLARE DEF accel
    LET xk1=tstep*v
    LET vk1=tstep*accel(x,v,t,w,g,q)
    LET xk2=tstep*(v+vk1/2)
    LET vk2=tstep*accel(x+xk1/2,v+vk1/2,t+tstep/2,w,g,q)
    LET xk3=tstep*(v+vk2/2)
    LET vk3=tstep*accel(x+xk2/2,v+vk2/2,t+tstep/2,w,g,q)
    LET xk4=tstep*(v+vk3)
    LET vk4=tstep*accel(x+xk3,v+vk3,t+tstep,w,g,q)
    LET vnew=v+(vk1+2*vk2+2*vk3+vk4)/6
    LET xnew=x+(xk1+2*xk2+2*xk3+xk4)/6
END SUB
```

Listing 5

```
!PROGRAM TITLE ***************BIFURCATION****************
CLEAR
PRINT"                   ***PENDULUM - BIFURCATION DIAGRAM***"
PRINT"THIS PROGRAM DISPLAYS THE BIFURCATION DIAGRAM FOR THE PENDULUM. "
```

```
PRINT"FOR EACH VALUE OF FORCING AMPLITUDE, G, THE SYSTEM COMES TO A "
PRINT"STEADY STATE (AFTER MIN.TIME) AND THEN THE ANGULAR VELOCITY AT"
PRINT"AT THE BEGINNING OF EACH FORCING CYCLE IS DISPLAYED FOR A NUMBER OF"
PRINT"FURTHER CYCLES (GOVERNED BY MAX. TIME).  THE DATA CAN BE SAVED TO A FILE"
PRINT

DIM XINT(10), VINT(10)
LIBRARY "SGLIB.TRC"
!
DECLARE DEF accel
DIM A(1),B(1)
INPUT prompt"Input LOWEST DRIVING FORCE STRENGTH: ":GMIN
INPUT PROMPT"INPUT HIGHEST DRIVING FORCE STRENGTH:":GMAX
INPUT PROMPT"INPUT G STEPSIZE:":DELTAG
INPUT PROMPT"INPUT NUMBER OF SETS OF INITIAL CONDITIONS:":NUMSETS
FOR I=1 TO NUMSETS
    INPUT PROMPT"INPUT INITIAL ANGLE:":XINT(i)
    INPUT PROMPT"INPUT INITIAL ANGULAR VELOCITY:":VINT(I)
NEXT I
INPUT prompt"Input damping (If no damping then input 9999999):":q
INPUT Prompt"Input min. and max. time:":tmin,tmax
INPUT prompt"Input phase angle/(2*pi) - USE ZERO IF WANT BEGINNING OF CYCLE:":PH
I
PRINT
PRINT"SINCE THE RUNTIME IS VERY LONG THE NEXT SET OF INPUTS GIVE AN OPTION"
PRINT"TO SAVE THE DATA TO A FILE"
INPUT PROMPT"SAVE TO A FILE? YES(1), NO(2):":SV
IF SV=1 THEN
    PRINT "A REASONABLE 8 CHARACTER FILE NAME (USE A NUMBER) MIGHT INCLUDE"
    PRINT"1)FIRST 2 DIGITS FOR Q VALUE"
    PRINT"2)NEXT 3 DIGITS FOR LOWEST G VALUE"
    PRINT"3)LAST 3 DIGITS FOR HIGHES G VALUE"
    PRINT" EXAMPLE   20145150"
    INPUT PROMPT"FILE NAME =>":FILENAME
    INPUT PROMPT"DATA FILE DRIVE (A/B/C/D):":B$
    LET NAME$=STR$(FILENAME)
END IF
CLEAR
CALL PARAMS(W,EPS,TSTEP)
CALL SETAXES(0)
CALL SETXSCALE(GMIN,GMAX)
CALL SETYSCALE(-1,3)
CALL SETTEXT("PENDULUM BIFURCATION DIAGRAM","FORCING-G","ANGULAR VELOCITY")
CALL RESERVELEGEND
!
DATA 0,0
CALL DATAGRAPH(A,B,1,0,"WHITE")
!
IF SV=1 THEN                          !OPENS A FILE AND SPECIFIES CHARACTERISTICS
    OPEN #1:NAME B$&":"&NAME$,ORGANIZATION RECORD, CREATE NEWOLD
    ASK #1:FILESIZE LENGTH
    IF LENGTH=0 THEN SET#1: RECSIZE 10
    SET #1: POINTER END
END IF
!
FOR II=1 TO NUMSETS                   !LOOPS FOR ALL INITIAL CONDITIONS
    LET T=0
    LET XP =XINT(ii)
    LET VP= VINT(II)
    FOR G=GMIN TO GMAX STEP DELTAG        !LOOPS FOR ALL G VALUES
        LET t=0
        LET x=xP
        LET v=vP
        CALL GOTOCANVAS
        !
        !CALCULATION AND GRAPHING BLOCK
        LET phi=phi*2*pi
        FOR i=1 to 1000000
            CALL rk4(x,v,tstep,xnew,vnew,t,w,g,q)      ! Take a step of size tste
p
            LET tshalf=tstep/2
            CALL rk4(x,v,tshalf,xnh,vnh,t,w,g,q)       !Take two half steps
            CALL rk4(xnh,vnh,tshalf,xn,vn,t+tshalf,w,g,q)
```

```
                    LET d1=abs(xn-xnew)
                    LET d2=abs(vn-vnew)
                    LET delta=max(d1,d2)
                    IF delta<eps then
                        IF t>tmin then
                            LET tnew=t+tstep
                            LET w1=mod(phi-w*t,2*pi)   !Check for Poincare section
                            LET w2=mod(w*tnew-phi,2*pi)
                            IF w1<w*tstep then
                                IF w2<w*tstep then
                                    LET ts=w1/w
                                    CALL rk4(x,v,ts,xp,vp,t,w,g,q)        !CALCULATES POINT AT
SECTION
                                    IF abs(xp)>pi then LET xp=xp-2*pi*abs(xp)/xp
                                    CALL GRAPHPOINT(G,VP,1)
                                    IF SV=1 THEN WRITE #1:G,VP
                                END IF
                            END IF
                        END IF
                        LET x=xnew
                        LET v=vnew
                        LET t=t+tstep        !Expand step size
                        LET tstep=tstep*.95*(eps/delta)^.25
                        IF abs(x)>pi then   !bring theta back into range
                            LET x=x-2*pi*abs(x)/x
                        END IF
                    ELSE                     !else reduce step size
                        LET tstep=tstep*.95*(eps/delta)^.2
                    END IF
                    IF t>tmax then LET i=1000001
                NEXT i
            NEXT G
NEXT II
LET G$=STR$(G)
LET Q$=STR$(Q)
CALL ADDLEGEND(" Q="&STR$(Q)&"    PHI="&STR$(PHI),0,1,"WHITE")
CALL DRAWLEGEND
get key variable
clear
print"press <esc> key to finish"
END
!
!
SUB rk4(x,v,tstep,xnew,vnew,t,w,g,q)
    DECLARE DEF accel
    LET xk1=tstep*v
    LET vk1=tstep*accel(x,v,t,w,g,q)
    LET xk2=tstep*(v+vk1/2)
    LET vk2=tstep*accel(x+xk1/2,v+vk1/2,t+tstep/2,w,g,q)
    LET xk3=tstep*(v+vk2/2)
    LET vk3=tstep*accel(x+xk2/2,v+vk2/2,t+tstep/2,w,g,q)
    LET xk4=tstep*(v+vk3)
    LET vk4=tstep*accel(x+xk3,v+vk3,t+tstep,w,g,q)
    LET vnew=v+(vk1+2*vk2+2*vk3+vk4)/6
    LET xnew=x+(xk1+2*xk2+2*xk3+xk4)/6
END SUB
DEF accel(x,v,t,w,g,q)
    LET accel=-sin(x)-(1/q)*v+g*cos(w*t)
END DEF
!
SUB PARAMS(W,EPS,TSTEP)
    LET W=0.66666666
    LET EPS=1.0E-6
    LET TSTEP=0.5
END SUB
```

Listing 6

```
1000 !PROGRAM TITLE ****************FFT*****************
1001 CLEAR
1002 PRINT"                    ***PENDULUM - FAST FOURIER TRANSFORM OF VARIABLES***"
1003 PRINT
1004 PRINT"THIS PROGRAM PROVIDES A PHASE DIAGRAM, TIME SERIES, AND FFT OF THE"
1005 PRINT"ANGLE OR ANG. VELOCITY OF THE PENDULUM.  A BLINKING CURSOR INDICATES"
1006 PRINT"THAT THE PROGRAM IS READY FOR THE NEXT STEP.  PRESS ANY KEY TO CONTNU
E."
1007 PRINT"THE HANNING OPTION IS USED TO SMOOTH THE EFFECT OF THE ABRUPT WINDOW"
1008 PRINT"AND IS RECOMMENDED IN MOST CASES."
1009 PRINT
1010 LIBRARY "SGLIB.TRC"
1020 !
1030 DIM thetadata(5000),thetadotdata(5000),xreal(0 to 5000),ximag(0 to 10000)
1040 DIM tpoint(0 to 5000),power(2048),frequency(2048)
1050 DECLARE DEF accel
1060 DECLARE DEF bitr
1070 INPUT prompt"Max frequency (try 0.5): ":maxfreq
1080 INPUT prompt"Input driving force strength: ":g
1090 INPUT prompt"Input damping term:":q
1100 INPUT prompt"Input initial position: ": xint
1110 INPUT prompt"Input initial velocity: ": vint
1120 INPUT Prompt"Input min.time:":tmin
1130 INPUT prompt"No. of FFT points(..256,512,1024,2048,4096) : ":number
1140 PRINT"Desired power spectrum quantity"
1150 PRINT"  1)Power spectrum of angular velocity"
1160 PRINT"  2)Power spectrum of angle"
1170 INPUT prompt"Choose 1 or 2 :":ps
1180
1190 LET del=.5/maxfreq
1200 LET tmax=number*del+tmin
1210 LET w=0.6666667
1220 LET eps=1.0e-6
1230 LET tstep=0.5
1240
1250 LET t=0
1260 LET x=xint
1270 LET v=vint
1280 LET sumvel=0
1290 LET count=0
1300 LET p=1
1310 CALL SETTEXT("PENDULUM PHASE DIAGRAM","ANGLE","ANGULAR VELOCITY")
1320 CALL RESERVELEGEND
1330 PRINT"computing data"
1340 !
1350 FOR i=1 to 10000000
1360     CALL rk4(x,v,tstep,xnew,vnew,t,w,g,q)    ! Take a step of size tstep
1370     LET tshalf=tstep/2
1380     CALL rk4(x,v,tshalf,xnh,vnh,t,w,g,q)      !Take two half steps
1390     CALL rk4(xnh,vnh,tshalf,xn,vn,t+tshalf,w,g,q)
1400     LET d1=abs(xn-xnew)
1410     LET d2=abs(vn-vnew)
1420     LET delta=max(d1,d2)
1430     IF delta<eps then
1440        IF t>tmin then
1450           LET tnew=t + tstep
1460           LET w1=mod(-t,del)
1470           LET w2=mod(tnew,del)
1480           IF w1<tstep then
1490              IF w2<tstep then
1500                 LET ts=w1
1510                 CALL rk4(x,v,ts,xp,vp,t,w,g,q)
1520                 IF abs(xp)>pi then LET xp=xp-2*pi*abs(x)/x
1530                 LET thetadata(p)=xp
1540                 LET thetadotdata(p)=vp
1550                 LET p=p+1
1560                 LET sumvel= sumvel + v
1570                 LET count = count + 1
```

```
1580              END IF
1590            END IF
1600        END IF
1610        LET x=xnew
1620        LET v=vnew
1630        LET t=t+tstep
1640        LET tstep=tstep*.95*(eps/delta)^.25
1650        IF abs(x)>pi then
1660            LET x=x-2*pi*abs(x)/x
1670        END IF
1680     ELSE
1690         LET tstep=tstep*.95*(eps/delta)^.2
1700      END IF
1710      IF t>tmax then LET i=10000001
1720 NEXT i
1730 LET n=p-1
1740 LET meanvel=sumvel/count
1750 CLEAR
1760 CALL setgraphtype("xy")
1770 CALL datagraph(thetadata,thetadotdata,1,0,"white")
1780 CALL ADDLEGEND("G="&STR$(G)&"    Q="&str$(Q),0,1,"WHITE")
1790 CALL DRAWLEGEND
1800 GET KEY keyvariable
1810 !
1820 !
1830 !PREPARATION OF THE FFT DATA
1840 CLEAR
1850 INPUT prompt" HANNING OPTION Y/N? ": hanning$
1860 LET tgamma=log2(n)
1870 IF abs(int(tgamma)-tgamma)=0 then
1880     LET gamma=tgamma
1890     GOTO 1920
1900 END IF
1910 LET gamma=int(tgamma)+1
1920 PRINT "gamma= ";gamma
.1930 LET newn=2^gamma
1940 LET nu=gamma
1950 FOR i=n+1 to newn
1960     LET xreal(i)=0
1970 NEXT i
1980 LET n=newn
1990 PRINT"n=",n
2000 CLEAR
2010 IF ps=1 then LET title$="ANGULAR VELOCITY"
2020 IF ps=2 then LET title$="ANGLE"
2030 CALL settext("TIME SERIES","TIME",title$)
2040 CALL setxscale(tmin,tmax)
2050 FOR k=0 to n-1
2060
2070      IF ps = 1 then
2080          LET xreal(k)=thetadotdata(k+1)
2090      ELSE IF ps = 2 then
2100          LET xreal(k)=thetadata(k+1)
2110      END IF
2120      IF hanning$="y" then LET xreal(k) = xreal(k)*(.5-.5*cos(2*pi*k/(n-1)))
2130      LET ximag(k)=0
2140      LET tpoint(k)=tmin+k*del
2150 NEXT k
2160 CALL setgraphtype("")
2170 CALL datagraph(tpoint,xreal,1,0,"white")
2180 GET KEY keyvariable
2190 FOR i= 1 to 100
2200 NEXT i
2210
2220 !FFT ALGORITHM
2230 CLEAR
2240 PRINT "Calculating FFT"
2250 LET n2=n/2
2260 LET nu1=nu-1
2270 LET k=0
2280 FOR l=1 to nu
2290     DO while k<(n-1)
2300         FOR i=1 to n2
```

```
2310          LET argument=k/2^nu1
2320          LET garbage=int(argument)
2330          LET p=bitr(garbage,nu)
2340          LET arg =2*pi*p/n
2350          LET c=cos(arg)
2360          LET s=sin(arg)
2370          LET k1=k+1
2380          LET k1n2=k1+n2
2390          LET treal=xreal(k1n2)*c+ximag(k1n2)*s
2400          LET timag=ximag(k1n2)*c-xreal(k1n2)*s
2410          LET xreal(k1n2)=xreal(k1)-treal
2420          LET ximag(k1n2)=ximag(k1)-timag
2430          LET xreal(k1)=xreal(k1)+treal
2440          LET ximag(k1)=ximag(k1)+timag
2450          LET k=k+1
2460       NEXT i
2470       LET k=k+n2
2480    LOOP
2490    LET k=0
2500    LET nu1=nu1-1
2510    LET n2=int(n2/2)
2520 NEXT l
2530
2540 FOR k=1 to n
2550    LET i=bitr(k-1,nu)+1
2560    IF i<=k then GOTO 2630
2570    LET treal=xreal(k)
2580    LET timag=ximag(k)
2590    LET xreal(k)=xreal(i)
2600    LET ximag(k)=ximag(i)
2610    LET xreal(i)=treal
2620    LET ximag(i)=timag
2630 NEXT k
2640
2650 !GRAPHING THE FFT
2660 CLEAR
2670 INPUT prompt"Plot 1)power spectrum, or 2)log power spectrum: ":pps
2680 INPUT prompt"Frequency variable - 1)linear, or 2)log: ":freqvar
2690 LET maxfreq=.5/del
2700 LET minfreq=1/(number*del)
2710
2720 CLEAR
2730 !Y-AXIS
2740 IF pps = 1 then
2750    LET TITLE$="POWER SPECTRUM"
2760    LET YAXIS$="POWER"
2770 ELSE
2780    LET TITLE$="LOG POWER SPECTRUM"
2790    LET YAXIS$="LOG POWER"
2800 END IF
2810 !X-AXIS
2820 IF freqvar=2 then
2830    LET XAXIS$="LOG FREQUENCY"
2840 ELSE
2850    LET XAXIS$="FREQUENCY"
2860 END IF
2870
2880 !DRAW AXES
2890 CLEAR
2900 CALL setxscale(minfreq,maxfreq)
2910 CALL setyscale(1e-6,.99)
2920 CALL SETTEXT(TITLE$,XAXIS$,YAXIS$)
2930 CALL RESERVELEGEND
2940
2950 !PLOT POINTS
2960 FOR i=1 to n/2
2970    LET frequency(i)=i/(n*del)
2980    LET power(i)=(((xreal(i))^2+(ximag(i))^2))/(n^2)
2990
3000 NEXT i
3010 !PLOT TEXT
3020 CALL setaxes(0)
3030 IF pps=1 then
```

```
3040    IF freqvar=1 then CALL setgraphtype("xy")
3050    IF freqvar=2 then CALL setgraphtype("logx")
3060 END IF
3070 IF pps=2 then
3080    IF freqvar=1 then CALL setgraphtype("logy")
3090    IF freqvar=2 then CALL setgraphtype("logxy")
3100 END IF
3102 IF NUMBER=4096 THEN
3110    CALL datagraph(frequency,power,1,1,"white")
3112 ELSE
3114    CALL DATAGRAPH(FREQUENCY,POWER,1,0,"WHITE")
3116 END IF
3120 CALL ADDLEGEND("N="&STR$(N)&"   MAX FREQ ="&STR$(MAXFREQ)&"   DEL F="&STR$(MI
NFREQ),0,1,"WHITE")
3130 CALL ADDLEGEND("  G="&STR$(G)&"   Q="&STR$(Q),0,1,"WHITE")
3140 CALL drawlegend
3150 IF hanning$="y" then
3160    !   CALL ADDLEGEND("  HANNING",0,1,"WHITE")
3170 END IF
3180 GET KEY keyvariable
3190 INPUT PROMPT "Another with Hanning? y/n: ":hann$
3200 IF hann$= "y" THEN GOTO 1840
3210 INPUT PROMPT "FFT of different quantity (y/n)? :": diffquant$
3220 IF diffquant$ = "y" THEN
3230    IF ps = 2 THEN LET ps = 1
3240    IF ps = 1 THEN LET ps = 2
3250    GOTO 2000
3260 END IF
3270 INPUT PROMPT "Different presentation of same FFT? (y/n): ":diffplot$
3280 IF diffplot$ ="y" THEN GOTO 2660
3290 END
3300 !
3310 SUB rk4(x,v,tstep,xnew,vnew,t,w,g,q)
3320    DECLARE DEF accel
3330    LET xk1=tstep*v
3340    LET vk1=tstep*accel(x,v,t,w,g,q)
3350    LET xk2=tstep*(v+vk1/2)
3360    LET vk2=tstep*accel(x+xk1/2,v+vk1/2,t+tstep/2,w,g,q)
3370    LET xk3=tstep*(v+vk2/2)
3380    LET vk3=tstep*accel(x+xk2/2,v+vk2/2,t+tstep/2,w,g,q)
3390    LET xk4=tstep*(v+vk3)
3400    LET vk4=tstep*accel(x+xk3,v+vk3,t+tstep,w,g,q)
3410    LET vnew=v+(vk1+2*vk2+2*vk3+vk4)/6
3420    LET xnew=x+(xk1+2*xk2+2*xk3+xk4)/6
3430 END SUB
3440 !
3450 !
3460 DEF accel(x,v,t,w,g,q)
3470    LET accel=-sin(x)-(1/q)*v+g*cos(w*t)
3480 END DEF
3490
3500 !BIT REVERSER FUNCTION
3510 DEF bitr(j,nu)
3520    LET j1=j
3530    LET ibitr=0
3540    FOR i=1 to nu
3550        LET j2 = int(j1/2)
3560        LET ibitr=ibitr*2+(j1-2*j2)
3570        LET j1=j2
3580    NEXT i
3590    LET bitr=ibitr
3600 END DEF
```

Listing 7

```
!PROGRAM TITLE ***********BASINS**************
CLEAR
PRINT"                      ***PENDULUM - BASINS OF ATTRACTION***"
PRINT"This program calculates the average angular velocity for"
PRINT"an array of initial conditions:"
PRINT"velocity (-3,3) and angle (-pi ,pi).  If the average angular velocity"
PRINT"from a given initial point is positive then circle is placed at the"
PRINT"point, otherwise the average is negative.  This program can also "
PRINT"superpose the appropriate Poincare section on the graph."
PRINT"The program can also save the data sets to two different files."
PRINT
LIBRARY "sglib.trc"
DIM a(1),b(1)

DECLARE DEF accel

INPUT prompt"Input driving force strength: ":g
INPUT prompt"input damping :":q
INPUT Prompt"Input min. and max. time of averaging:":tmin,tmax
INPUT Prompt"Poincare attractor yes (1), no(2) ":Poincare
INPUT Prompt"Save data to a file, yes(1), no(2):":sv
IF sv=1 then
    PRINT"Basin of attraction File name includes:"
    PRINT"  1)First 2 digits for q value"
    PRINT"  2)Next needed digits for g value"
    PRINT"  3)Last digits as 0's"
    INPUT prompt"file name  - format ex. 20150000:":filename
    INPUT prompt"data file drive a,b,c,etc.?":b$
    IF poincare=1 then
        PRINT"Superposed Poincare section file name is similar to above except"
        PRINT"that the numeral '0' is added, for example, 20015000."
        INPUT prompt"Input corresponding Poincare file name:":poinfile
    END IF
    LET FILENAME$=STR$(FILENAME)
    LET POINFILE$=STR$(POINFILE)
END IF
!
CALL PARAMS(W,EPS,TSTEP)
CALL SETXSCALE(-3,3)
CALL SETYSCALE(-3,3)
CALL SETTEXT("PENDULUM - BASINS OF ATTRACTION","INIT. ANGLE","INIT. ANG. VEL.")
CALL RESERVELEGEND
IF sv=1 then
    OPEN #1:name b$&":"&filename$,organization record,create newold
    ASK #1: FILESIZE length
    IF length=0 then SET#1:rECSIZE 10
    SET #1: POINTER end
    IF poincare=1 then
        OPEN #2:name b$&":"&poinfile$,organization record,create newold
        ASK #2: FILESIZE length
        IF length=0 then SET#2:rECSIZE 10
        SET #2:pOINTER end
    END IF
END IF

DATA 0,0
CALL DATAGRAPH(A,B,1,0,"WHITE")
CALL gotocanvas
LET phi=0
FOR xint=-pi to pi step .1
    FOR vint=-3 to 3 step .15
        LET x= xint
        LET v=vint
        LET t=0
        LET s=0
        FOR i=1 to 1000000
            CALL rk4(x,v,tstep,xnew,vnew,t,w,g,q)        ! Take a step of size tste
```

```
p
                LET tshalf=tstep/2
                CALL rk4(x,v,tshalf,xnh,vnh,t,w,g,q)          !Take two half steps
                CALL rk4(xnh,vnh,tshalf,xn,vn,t+tshalf,w,g,q)
                LET d1=abs(xn-xnew)
                LET d2=abs(vn-vnew)
                LET delta=max(d1,d2)
                IF delta<eps then
                    IF t>tmin then
                        LET tnew=t+tstep
                        LET s=s+vnew*tstep
                        IF Poincare=1 then
                            LET w1=mod(phi-w*t,2*pi)
                            LET w2=mod(w*tnew-phi,2*pi)
                            IF w1<w*tstep then
                                IF w2<w*tstep then
                                    LET ts=w1/w
                                    CALL rk4(x,v,ts,xp,vp,t,w,g,q)
                                    IF abs(xp)>pi then LET xp=xp-2*pi*abs(xp)/xp
                                    CALL GRAPHPOINT(xp,vp,1)
                                    IF sv=1 then WRITE #2:xp,vp
                                END IF
                            END IF
                        END IF
                    END IF
                    LET x=xnew
                    LET v=vnew
                    LET t=t+tstep        !Expand step size
                    LET tstep=tstep*.95*(eps/delta)^.25
                    IF abs(x)>pi then    !bring theta back in range
                        LET x=x-2*pi*abs(x)/x
                    END IF
                ELSE                     !else reduce step size
                    LET tstep=tstep*.95*(eps/delta)^.2
                END IF
                IF t>tmax then
                    LET average=s/(tmax-tmin)
                    IF average>0 then
                        CALL GRAPHPOINT(XINT,VINT,4)
                        IF sv=1 then WRITE #1:xint,vint
                    END IF
                    LET i=1000001
                END IF
            NEXT i
        NEXT vint
NEXT xint
CALL addlegend("g="&str$(q)&"   q="&str$(q)&"   circle=positive",0,1,"white")
CALL drawlegend
GET KEY VARIABLE
CLEAR
print"press <esc> key to finish"
END
!
!
SUB rk4(x,v,tstep,xnew,vnew,t,w,g,q)
    DECLARE DEF accel
    LET xk1=tstep*v
    LET vk1=tstep*accel(x,v,t,w,g,q)
    LET xk2=tstep*(v+vk1/2)
    LET vk2=tstep*accel(x+xk1/2,v+vk1/2,t+tstep/2,w,g,q)
    LET xk3=tstep*(v+vk2/2)
    LET vk3=tstep*accel(x+xk2/2,v+vk2/2,t+tstep/2,w,g,q)
    LET xk4=tstep*(v+vk3)
    LET vk4=tstep*accel(x+xk3,v+vk3,t+tstep,w,g,q)
    LET vnew=v+(vk1+2*vk2+2*vk3+vk4)/6
    LET xnew=x+(xk1+2*xk2+2*xk3+xk4)/6
END SUB
!
!
DEF accel(x,v,t,w,g,q)
    LET accel=-sin(x)-(1/q)*v+g*cos(w*t)
END DEF
!
SUB PARAMS(W,EPS,TSTEP)
    LET W=.6666666666
    LET EPS=1E-6
    LET TSTEP=.5
END SUB
```

Listing 8

```
!PROGRAM TITLE***************LOGISTIC MAP************
LIBRARY "SGLIB.TRC"
DECLARE DEF LOGISTIC
DIM CELL(1000),PROB(1000),G(1),H(1)
CLEAR
PRINT"                 COMPREHENSIVE LOGISTIC MAP PROGRAM"
PRINT
PRINT"CHOOSE ONE OF THE FOLLOWING OPTIONS FOR THE LOGISTIC MAP:"
PRINT" 1)RETURN MAP"
PRINT" 2)BIFURCATION DIAGRAM"
PRINT" 3)ENTROPY DIAGRAM"
PRINT" 4)LYAPUNOV EXPONENT DIAGRAM"
INPUT PROMPT" CHOOSE 1,2,3, OR 4: ":CHOICE
CLEAR
!
CALL LINPUTS(CHOICE,XINT,LAMBDA,INITLAMBDA,FINLAMBDA,STEPLAMBDA,INITNUM,FINNUM,O
RDER,NUMCELLS,XMIN,XMAX,YMIN,YMAX,TITLE1$,TITLE2$,VLABEL$,HLABEL$)
!
!GRAPHING SET-UP PROCEDURE
CALL SETXSCALE(XMIN,XMAX)
CALL SETYSCALE(YMIN,YMAX)
CALL SETAXES(0)
CALL SETTEXT(TITLE1$,HLABEL$,VLABEL$)
CALL RESERVELEGEND
DATA 0,0
CALL DATAGRAPH(G,H,0,0,"WHITE")
CALL GOTOCANVAS
!
IF CHOICE=1 THEN CALL CALCULATION1(LAMBDA,XINT,INITNUM,FINNUM,ORDER)
IF CHOICE=2 THEN CALL CALCULATION2(XINT,INITLAMBDA,FINLAMBDA,STEPLAMBDA,INITNUM,
FINNUM)
IF CHOICE=3 THEN CALL CALCULATION3(XINT,INITLAMBDA,FINLAMBDA,STEPLAMBDA,INITNUM,
FINNUM,NUMCELLS)
IF CHOICE=4 THEN CALL CALCULATION4(XINT,INITLAMBDA,FINLAMBDA,STEPLAMBDA,INITNUM,
FINNUM)
!
SUB LINPUTS(CHOICE,XINT,LAMBDA,INITLAMBDA,FINLAMBDA,STEPLAMBDA,INITNUM,FINNUM,OR
DER,NUMCELLS,XMIN,XMAX,YMIN,YMAX,TITLE1$,TITLE2$,VLABEL$,HLABEL$)
    INPUT PROMPT"INPUT INITIAL X VALUE:":XINT
    IF CHOICE=1 THEN INPUT PROMPT"INPUT LAMBDA:":LAMBDA
    INPUT PROMPT"INPUT NUMBER OF INITIAL THROWAWAY ITERATIONS:":INITNUM
    INPUT PROMPT"INPUT TOTAL NUMBER OF ITERATIONS:":FINNUM
    IF CHOICE>1 THEN
        INPUT PROMPT"INPUT LOWEST LAMBDA VALUE:":INITLAMBDA
        INPUT PROMPT"INPUT HIGHEST LAMBDA VALUE:":FINLAMBDA
    END IF
    IF CHOICE=1 THEN INPUT PROMPT"INPUT ORDER OF MAP:":ORDER
    IF CHOICE=3 THEN INPUT PROMPT"INPUT # OF CELLS:":NUMCELLS
    IF CHOICE=1 THEN LET ORDER$=STR$(ORDER)
    IF CHOICE=1 THEN
        LET XMIN=0
        LET XMAX=1
        LET YMIN=0
        LET YMAX=1
        LET TITLE1$="LOGISTIC MAP"
        LET TITLE2$="X(N+"&ORDER$&")   VERSUS X(N)"
        LET VLABEL$="N+"&ORDER$&" VALUE"
        LET HLABEL$="N VALUE"
    END IF
    IF CHOICE>1 THEN
        LET XMIN=INITLAMBDA
        LET XMAX=FINLAMBDA
        LET STEPLAMBDA=(XMAX-XMIN)/740
    END IF
    IF CHOICE=2 THEN
        LET YMIN=0
        LET YMAX=1
        LET TITLE1$="LOGISTIC MAP BIFURCATION DIAGRAM"
```

```
          LET TITLE2$="LAMBDA="&STR$(XMIN)&" TO LAMBDA="&STR$(XMAX)
          LET VLABEL$="X"
          LET HLABEL$="LAMBDA"
       END IF
       IF CHOICE=3 THEN
          LET YMIN=INT(LOG(1/NUMCELLS))
          LET YMAX=1
          LET TITLE1$="LOGISTIC MAP ENTROPY"
          LET TITLE2$="LAMBDA="&STR$(INITLAMBDA)&" TO LAMBDA="&STR$(FINLAMBDA)&":
 #CELLS="&STR$(NUMCELLS)&":     RANDOM ENTROPY="&STR$(-LOG(NUMCELLS))
          LET VLABEL$="X"
          LET HLABEL$="LAMBDA"
       END IF
       IF CHOICE=4 THEN
          LET YMIN=-3
          LET YMAX=1
          LET TITLE1$="LOGISTIC MAP LYAPUNOV EXPONENTS"
          LET TITLE2$="LAMBDA="&STR$(INITLAMBDA)&" TO LAMBDA="&STR$(FINLAMBDA)
          LET VLABEL$="LYAP EXP"
          LET HLABEL$="LAMBDA"
       END IF
END SUB

SUB CALCULATION1(LAMBDA,XINT,INITITERATION,ITERATIONNUM,ORDER)
      PLOT LINES: 0,0;1,1            'PLOTS X(N+1)=X(N) LINE
      FOR I=0 TO 1 STEP .001         'PLOTS CURVE OF Y=LAMBDA*X*(1-X)
          LET J=I
          FOR ORD=1 TO ORDER
              LET Z=LAMBDA*J*(1-J)
              LET J=Z
          NEXT ORD
          PLOT I,Z
      NEXT I
      LET X=XINT                     'PLOTS RETURN MAP
      IF INITITERATION= 0 THEN
         PLOT X,0;
      END IF
      FOR I=1 TO ITERATIONNUM
          LET J=X
          FOR ORD= 1 TO ORDER
              LET Y=LOGISTIC(J,LAMBDA)
              LET J=Y
          NEXT ORD
          IF I>=INITITERATION THEN
             PLOT X,Y;
             PLOT Y,Y;
          END IF
          LET X=Y
      NEXT I
END SUB
!
SUB CALCULATION2(XINT,INITLAMBDA,FINLAMBDA,STEPLAMBDA,INITNUM,FINNUM)
      FOR LAMBDA=INITLAMBDA TO FINLAMBDA STEP STEPLAMBDA
          LET X=XINT
          FOR I=1 TO FINNUM
              LET Y=LOGISTIC(X,LAMBDA)
              IF I>INITNUM THEN
                 PLOT LAMBDA,Y
              END IF
              LET X=Y
          NEXT I
      NEXT LAMBDA
END SUB
!
SUB CALCULATION3(XINT,INITLAMBDA,FINLAMBDA,STEPLAMBDA,INITNUM,FINNUM,NUMCELLS)
      FOR LAMBDA=INITLAMBDA TO FINLAMBDA STEP STEPLAMBDA
          LET ENTROPY=0
          FOR L=1 TO NUMCELLS
              LET CELL(L)=0
          NEXT L
          LET X=XINT
          FOR I=1 TO FINNUM
```

```
                LET Y=LOGISTIC(X,LAMBDA)
                IF I>INITNUM THEN
                    LET L=INT(NUMCELLS*Y)+1
                    LET CELL(L)=CELL(L)+1
                END IF
                LET X=Y
            NEXT I
            FOR L=1 TO NUMCELLS
                LET PROB(L)=CELL(L)/(FINNUM-INITNUM)
                IF PROB(L)>O THEN
                    LET ENTROPY=ENTROPY+PROB(L)*LOG(PROB(L))
                END IF
            NEXT L
            PLOT LAMBDA,ENTROPY;
        NEXT LAMBDA
END SUB
!
SUB CALCULATION4(XINT,INITLAMBDA,FINLAMBDA,STEPLAMBDA,INITNUM,FINNUM)
    FOR LAMBDA=INITLAMBDA TO FINLAMBDA STEP STEPLAMBDA
        LET LYAP=O
        LET X=XINT
        FOR I = 1 TO FINNUM
            LET Y=LOGISTIC(X,LAMBDA)
            IF I>INITNUM THEN
                LET LYAP=LYAP+LOG(ABS(LAMBDA*(1-2*Y)))        !SUM LOG DERIVATIVES
            END IF
            LET X=Y
        NEXT I
        LET LYAP=LYAP/(FINNUM-INITNUM)
        IF LYAP<-3 THEN LET LYAP=-3
        PLOT LAMBDA,LYAP;
    NEXT LAMBDA
END SUB


DEF LOGISTIC(X,LAMBDA)
    LET LOGISTIC =LAMBDA*X*(1-X)
END DEF
CALL ADDLEGEND(TITLE2$,O,1,"WHITE")
CALL DRAWLEGEND
GET KEY VARIABLE
CLEAR
PRINT"PRESS <ESC> TO RETURN TO MENU"
END
```

Listing 9

```
!PROGRAM TITLE******************CIRCLE MAP*****************
LIBRARY "SGLIB.TRC"

DECLARE DEF CIRCLE
DIM XINT(10),G(1),H(1)
CLEAR
PRINT"                    COMPREHENSIVE CIRCLE MAP PROGRAM"
PRINT
PRINT"CHOOSE ONE OF THE FOLLOWING OPTIONS FOR THE CIRCLE MAP:"
PRINT" 1)RETURN MAP"
PRINT" 2)BIFURCATION MAP"
PRINT" 3)DEVIL'S STAIRCASE"
INPUT PROMPT" CHOOSE 1,2,OR 3:":CHOICE
CLEAR
!
CALL LINPUTS(CHOICE,XIN,XINT(),NUMXINT,KVALUE,INITK,FINK,STEPK,OMEGA,INITNUM,FIN
NUM,ORDER,XMIN,XMAX,YMIN,YMAX,TITLE1$,TITLE2$,VLABEL$,HLABEL$)
!
```

```
!GRAPHING SET-UP PROCEDURE
CALL SETXSCALE(XMIN,XMAX)
CALL SETYSCALE(YMIN,YMAX)
CALL SETAXES(0)
CALL SETTEXT(TITLE1$,HLABEL$,VLABEL$)
CALL RESERVELEGEND
DATA 0,0
CALL DATAGRAPH(G,H,0,0,"WHITE")
CALL GOTOCANVAS
!
IF CHOICE= 1 THEN CALL CALCULATION1(XIN,KVALUE,OMEGA,INITNUM,FINNUM,ORDER)
IF CHOICE= 2 THEN CALL CALCULATION2(NUMXINT,XINT,INITK,FINK,STEPK,OMEGA,INITNUM,
FINNUM)
IF CHOICE= 3 THEN CALL CALCULATION3(XIN,INITNUM,FINNUM,INITK,FINK,STEPK,KVALUE)
!
SUB LINPUTS(CHOICE,XIN,XINT(),NUMXINT,KVALUE,INITK,FINK,STEPK,OMEGA,INITNUM,FINN
UM,ORDER,XMIN,XMAX,YMIN,YMAX,TITLE1$,TITLE2$,VLABEL$,HLABEL$)
    IF CHOICE=2 THEN
        INPUT PROMPT"INPUT NUMBER OF INITIAL CONDITIONS:":NUMXINT
        FOR J = 1 TO NUMXINT
            INPUT PROMPT"INPUT AN INITIAL CONDITION:":XINT(J)
        NEXT J
    ELSE
        INPUT PROMPT"INPUT INITIAL ANGLE, [0,1] :":XIN
    END IF
    IF CHOICE = 3 THEN
        INPUT PROMPT"INPUT INITIAL VALUE OF OMEGA:":INITK
        INPUT PROMPT"INPUT FINAL VALUE OF OMEGA:":FINK
        LET STEPK=(FINK-INITK)/740
        INPUT PROMPT"INPUT K VALUE:":KVALUE
    ELSE IF CHOICE=2 THEN
        INPUT PROMPT"INPUT INITIAL VALUE OF K:":INITK
        INPUT PROMPT"INPUT FINAL VALUE OF K:":FINK
        LET STEPK=(FINK-INITK)/740
        INPUT PROMPT"INPUT OMEGA VALUE:":OMEGA
    END IF
    INPUT PROMPT"INPUT NUMBER OF THROWAWAY ITERATIONS:":INITNUM
    INPUT PROMPT"INPUT TOTAL NUMBER OF ITERATIONS:":FINNUM
    IF CHOICE = 1 THEN
        INPUT PROMPT"INPUT ORDER OF MAP:":ORDER
        INPUT PROMPT"INPUT OMEGA VALUE:":OMEGA
        INPUT PROMPT"INPUT K-VALUE:":KVALUE
        LET XMIN=0
        LET XMAX=1
        LET YMIN=0
        LET YMAX=1
        LET TITLE1$="CIRCLE MAP"
        LET TITLE2$="K="&STR$(KVALUE)&"   OMEGA="&STR$(OMEGA)
        LET VLABEL$="N+"&STR$(ORDER)&" ANGLE VALUE"
        LET HLABEL$="N ANGLE VALUE"
    END IF
    IF CHOICE = 2 THEN
        LET XMIN=INITK
        LET XMAX=FINK
        LET YMIN=0
        LET YMAX=1
        LET TITLE1$="CIRCLE MAP BIFURCATION DIAGRAM"
        LET TITLE2$=""
        LET VLABEL$="THETA"
        LET HLABEL$="K-VALUE"
    END IF
    IF CHOICE=3 THEN
        LET XMIN=INITK
        LET XMAX=FINK
        INPUT PROMPT"INPUT YMIN (USUALLY 0), AND YMAX (USUALLY 1):":YMIN,YMAX
        LET TITLE1$="DEVIL'S STAIRCASE  (CIRCLE MAP)"
        LET TITLE2$="DRIVE K="&STR$(KVALUE)
        LET VLABEL$="WINDING #"
        LET HLABEL$="OMEGA"
    END IF
END SUB
!
```

```
SUB CALCULATION1(XIN,KVALUE,OMEGA,INITNUM,FINNUM,ORDER)
    PLOT LINES: 0,0;1,1            !PLOTS DIAGONAL
    FOR I=0 TO 1 STEP .001        !PLOTS CURVE
        LET J=I
        FOR ORD=1 TO ORDER

            LET Z=CIRCLE(J,KVALUE,OMEGA)
            LET J=Z
        NEXT ORD
        PLOT I,Z
    NEXT I
    LET X=XIN
    LET Y=0
    IF INITNUM=0 THEN PLOT X,0;
    FOR I=1 TO FINNUM
        LET J=x
        FOR ORD=1 TO ORDER
            LET Y=CIRCLE(J,KVALUE,OMEGA)
            LET J=Y
        NEXT ORD
        IF I>=INITNUM THEN
            PLOT X,Y;
            PLOT Y,Y;
        END IF
        LET X=Y
    NEXT I
END SUB

SUB CALCULATION2(NUMXINT,XINT(),INITK,FINK,STEPK,OMEGA,INITNUM,FINNUM)
    FOR KVAL=INITK TO FINK STEP STEPK
        FOR K=1 TO NUMXINT
            LET X=XINT(K)
            FOR I=1 TO FINNUM
                LET Y=CIRCLE(X,KVAL,OMEGA)
                IF I>INITNUM THEN
                    PLOT KVAL,Y
                END IF
                LET X=Y
            NEXT I
        NEXT K
    NEXT KVAL
END SUB

SUB CALCULATION3(XIN,INITNUM,FINNUM,INITOMEGA,FINOMEGA,STEPOMEGA,KVALUE)
    FOR OMEGA=INITOMEGA TO FINOMEGA STEP STEPOMEGA
        LET SUM=0
        LET X=XIN
        FOR I=1 TO FINNUM
            LET Y=X+OMEGA-(KVALUE/(2*PI))*SIN(2*PI*X)
            IF I=INITNUM THEN LET X0=Y
            LET X=Y
        NEXT I
        LET WINDING=(Y-X0)/(FINNUM-INITNUM)
        IF WINDING <=YMAX THEN
            IF WINDING >=YMIN THEN
                PLOT OMEGA,WINDING
            END IF
        END IF
    NEXT OMEGA
END SUB

DEF CIRCLE(X,KVALUE,OMEGA)
    LET TEMPCIRCLE = X+OMEGA-(KVALUE/(2*PI))*SIN(2*PI*X)
    LET CIRCLE = MOD(TEMPCIRCLE,1)
END DEF
GET KEY VARIABLE
CLEAR
PRINT "PRESS <ESC> FOR MENU"
END
```

Listing 10

```
!PROGRAM NAME******************************HENON*************************
LIBRARY "SGLIB.TRC"
CLEAR
PRINT"                              ***HENON MAP***"
PRINT
PRINT"THIS PROGRAM GENERATES THE (X,Y) PHASE DIAGRAM FOR THE HENON MAP."
PRINT "        X(N+1)=1-A*X(N)^2 + Y(N)"
PRINT "        Y(N+1)=B*X(N)"
PRINT" TWO PARAMETERS ARE REQUIRED, A AND B. IF B=1 THE MAP IS "
PRINT"CONSERVATIVE.  IF B < ABS(1) THEN THE MAP IS DISSIPATIVE."
PRINT"TRY A=1.4 AND B=0.3 INITIALLY."
PRINT
!
DIM L(1),M(1)
INPUT PROMPT"INPUT INITIAL X , Y VALUES:":XINT,YINT
INPUT PROMPT"INPUT A,B VALUES:":A,B
INPUT PROMPT"INPUT NUMBER OF THROWAWAY ITERATIONS:":INITNUM
INPUT PROMPT"INPUT NUMBER OF TOTAL ITERATIONS:":FINNUM
!
CALL SETXSCALE(-1.5,1.5)
CALL SETYSCALE(-.5,.5)
CALL SETTEXT("HENON MAP","X","Y")
CALL RESERVELEGEND
!
DATA 0,0
CALL DATAGRAPH(L,M,1,0,"WHITE")
CALL GOTOCANVAS
LET X=XINT
LET Y=YINT
FOR I=INITNUM+1 TO FINNUM+1
    LET XNEW=1-A*(X^2)+Y
    LET YNEW=B*X
    CALL GRAPHPOINT(XNEW,YNEW,1)
    LET X=XNEW
    LET Y=YNEW
NEXT I
CALL ADDLEGEND("A="&STR$(A)&"    B="&STR$(B),0,1,"WHITE")
CALL DRAWLEGEND
get key variable
clear
print"press <esc> key to finish"
END
```

Listing 11

```
!PROGRAM TITLE        ****************PENDLYAP***************
LIBRARY "SGLIB.TRC"
DIM A(1),B(1)
CLEAR
PRINT"                    ***PENDULUM - LYAPUNOV EXPONENTS***"
PRINT
PRINT"THIS PROGRAM CALCULATES THE 3 LYAPUNOV EXPONENTS FOR THE DRIVEN PENDULUM"
PRINT"USING THE ALGORITHM OF WOLF ET AL.  THE EXPONENTS ARE SMOOTHED OVER "
PRINT"MANY DRIVE CYCLES (ORBITS)."
!
!N=# OF NONLINEAR EQUTNS., NN= TOTAL # OF EQUATIONS
LET N=3
LET NN=12
DECLARE DEF YPRIM
!
DIM Y(12), ZNORM(3), GSC(3), CUM(3), YNEW(12)
!
```

```
!INITIAL CONDITIONS FOR NONLINEAR SYSTEM
LET Y(1)=.5
LET Y(2)=0
LET Y(3)=0
!
!INITIAL CONDITIONS FOR LINEAR SYSTEM (ORTHONORMAL FRAME)
FOR  I= N+1 TO NN
     LET Y(I)=0.0
NEXT I
FOR I=1 TO n
     LET Y((N+1)*I) = 1.0
     LET CUM(I)=0.0
NEXT I
!
INPUT PROMPT"INTEGRATION STEPSIZE:":TSTEP
INPUT PROMPT"NUMBER OF ORBITS:":NUMORBITS
INPUT PROMPT"DRIVING FORCE:":G
INPUT PROMPT"DRIVE FREQUENCY:":W
INPUT PROMPT"DAMPING FACTOR:":Q
INPUT PROMPT"LOG BASE (1)NATURAL (2)INFO-2 CHOOSE 1 OR 2:":BASE
!
!SET UP GRAPHICS
IF BASE=1 THEN CALL SETYSCALE(-.7,.3)
IF BASE=2 THEN CALL SETYSCALE(-1,.5)
CALL SETXSCALE(0,NUMORBITS)
CALL SETTEXT("PENDULUM - LYAPUNOV EXPONENTS","# DRIVE CYCLES","LYAP. EXPS.")
CALL SETAXES(0)
CALL RESERVELEGEND
DATA 0,0
CALL DATAGRAPH(A,B,1,0,"WHITE")
CALL GOTOCANVAS
!
DO WHILE Y(3)<2*PI*NUMORBITS
    !INITIALIZE INTEGRATOR
    CALL  RKK4(Y(),NN,TSTEP,Q,W,G,YNEW())
    FOR K=1 TO 12
        LET Y(K)=YNEW(K)
    NEXT K
    !
    !CONSTRUCT A NEW ORTHONORMAL BASIS BY GRAM-SCHMIDT METHOD
    !NORMALIZE FIRST VECTOR
    LET ZNORM(1)=0.0
    FOR J = 1 TO N
        LET ZNORM(1) = ZNORM(1) +Y(N*J+1)^2
    NEXT J
    LET ZNORM(1)=(ZNORM(1))^.5
    FOR J=1 TO N
        LET Y(N*J+1)=Y(N*J+1)/ZNORM(1)
    NEXT J
    !
    !GENERATE THE NEW ORTHONORMAL SET OF VECTORS
    FOR J=2 TO N
        !  GENERATE J-1 COEFFICIENTS
        FOR K = 1 TO (J-1)
            LET GSC(K)=0.0
            FOR L = 1 TO N
                LET GSC(K) = GSC(K) + Y(N*L+J)*Y(N*L+K)
            NEXT L
        NEXT K
        !
        ! CONSTRUCT A NEW VECTOR
        FOR K = 1 TO N
            FOR L= 1 TO J-1
                LET Y(N*K+J) = Y(N*K+J) -GSC(L)*Y(N*K+L)
            NEXT L
        NEXT K
        !
        ! CALCULATE THE VECTOR'S NORM
        LET ZNORM(J) = 0.0
        FOR K= 1 TO N
            LET ZNORM(J) = ZNORM(J)+Y(N*K+J)^2
        NEXT K
        LET ZNORM(J) = (ZNORM(J))^.5
        !
        !
        ! NORMALIZE THE NEW VECTOR
```

```
            FOR K = 1 TO N
                LET Y(N*K+J) = Y(N*K+J)/ZNORM(J)
            NEXT K
        NEXT J
        !
        ! UPDATE RUNNING VECTOR MAGNITUDES
        FOR K = 1 TO N
            LET CUM(K) = CUM(K) +LOG(ZNORM(K))
            IF BASE = 2 THEN LET CUM(K) = CUM(K)/LOG(2)
        NEXT K
        !
        !NORMALIZE EXPONENT AND PLOT EXPONENTS
        IF Y(3)>0 THEN
            LET T=Y(3)/W
            FOR K = 1 TO N
                LET CUMKT=CUM(K)/T
                CALL GRAPHPOINT(T*W/(2*PI),CUMKT,1)
            NEXT K
        END IF
LOOP
!
CALL ADDLEGEND("G="&STR$(G)&"   Q="&STR$(Q)&"   W="&STR$(W),0,1,"WHITE")
CALL DRAWLEGEND
END
!
SUB RKK4(Y(),NN,TSTEP,Q,W,G,YNEW())
    DIM Y1(12), Y2(12), Y3(12), Y4(12), YY1(12), YY2(12), YY3(12)
    DECLARE DEF YPRIM
    FOR K= 1 TO NN
        LET Y1(K)=TSTEP*YPRIM(Y(),K,Q,W,G)
    NEXT K
    FOR K= 1 TO NN
        LET YY1(K)=Y(K)+Y1(K)/2
    NEXT K
    FOR K=1 TO NN
        LET Y2(K)=TSTEP*YPRIM(YY1(),K,Q,W,G)
    NEXT K
    FOR K = 1 TO NN
        LET YY2(K)=Y(K)+Y2(K)/2
    NEXT K
    FOR K = 1 TO NN
        LET Y3(K)=TSTEP*YPRIM(YY2(),K,Q,W,G)
    NEXT K
    FOR K = 1 TO NN
        LET YY3(K)=Y(K)+Y3(K)
    NEXT K
    FOR K= 1 TO NN
        LET Y4(K)=TSTEP*YPRIM(YY3(),K,Q,W,G)
    NEXT K
    FOR K = 1 TO NN
        LET YNEW(K) = Y(K)+(Y1(K) +2*Y2(K)+2*Y3(K)+Y4(K))/6
    NEXT K
END SUB


!DEFINE DERIVATIVES FUNCTIONS
DEF YPRIM(Y(),K,Q,W,G)
    IF K=1 THEN LET YPRIM=-Y(1)/Q-SIN(Y(2))+G*COS(Y(3))
    IF K=2 THEN LET YPRIM=Y(1)
    IF K=3 THEN LET YPRIM=W
    IF K>3 THEN
        IF K<7 THEN
            LET I = K-4
            LET YPRIM=-Y(4+I)/Q-Y(7+I)*COS(Y(2))+G*Y(10+I)*COS(Y(3))
        END IF
    END IF
    IF K>6 THEN
        IF K<10 THEN
            LET I = K-7
            LET YPRIM=Y(4+I)
        END IF
    END IF
    IF K>9  THEN LET YPRIM = 0
END DEF
```

References

Abraham, R.H., and Shaw, C.D. (1984). *Dynamics: the geometry of behaviour*, Part III, Ariel Press, Santa Cruz.

Abraham, N.B., Arimondo, E., and Boyd, R.W. (1988). Dynamics and chaos in nonlinear optical systems, in *Instabilities and chaos in quantum optics*, II, eds N.B. Abraham, F.T. Arecchi, and L.A. Lugiato, Plenum Pub. Corp., NY, pp. 375–91.

Arneodo, A., Grasseau, G., and Kostelich, E.J. (1987). Fractal dimensions and $f(\alpha)$ spectrum of the Hénon attractor, *Phys Lett. A*, **1254**, 426–32.

Atmanspacher, H., and Scheingraber, H. (1987). A fundamental link between system theory and statistical mechanics, *Foundations in Physics*, **17**, 939–63.

Atmanspacher, H., Scheingraber, H., and Voges, W. (1988). Global scaling properties of a chaotic attractor reconstructed from experimental data, *Phys. Rev. A.*, **37**, 1314–22.

Baierlein, R. (1971). *Atoms and information theory*, W.H. Freeman and Co., San Francisco, Chapter 3.

Bak, P. (1986). The Devil's staircase, *Physics Today*, December, 38–45.

Baker, G.L. (1986). A simple model of irreversibility, *Am. J. Phys.*, **54**, 704–8.

Bergé, F., Pomeau, Y., and Vidal, Ch. (1984). *Order within chaos*, John Wiley and Sons, NY, p. 179.

Blackburn, J.A., Vik, S., Binruo Wu, and Smith, H.J.T. (1989). Driven pendulum for studying chaos, *Rev. Sci. Instrum.*, **60**, 422–6.

Crutchfield, J.P., Farmer, J.D., Packard, N.H., and Shaw, R.S. (1986). Chaos, *Sci. Am.*, December, 46–57.

Davidson, A., Dueholm, B., and Beasley, M.R. (1986). Experiments on intrinsic and thermally induced chaos in an RF-driven Josephson junction, *Phys. Rev. B*, **33**, 5127–30.

D'Humieres, D., Beasley, M.R., Huberman, B.A., and Libchaber, A. (1982). Chaotic states and routes to chaos in the forced pendulum, *Phys. Rev. A*, **26**, 3483–96.

Dugas, R. (1958). *Mechanics in the 17th century*, Central Book Co., NY, p. 76.

Farmer, J.D., Ott, E., and Yorke, J.A. (1983). The dimension of chaotic attractors, *Physica*, **7D**, 153–80.

Feigenbaum, M.J. (1978). Quantitative universality for a class of nonlinear transformations, *J. Stat. Phys.*, **19**, 25–32.

Firth, W.J. (1986). Instabilities and chaos in lasers and optical resonators, in *Chaos*, ed. A.V. Holden, Princeton Univ. Press, Princeton, pp. 135–57.

Gioggia, R.S., and Abraham, N.B. (1983). Routes to chaotic output from a single-mode dc-excited laser, *Phys. Rev. Lett.*, **51**, 650–3.

Glazier, J.A., and Libchaber, A. (1988). Quasiperiodicity and dynamical systems: an experimentalist's view, *IEEE Trans. on Circuits and Systems*, **35**, No. 7, 790–809.

Gleick, J. (1987). *Chaos: making a new science*, Viking, NY.

Gollub, J.P., and Benson, S.V. (1980). Many routes to turbulent convection, *J. Fluid Mech.*, **100**, 449–70.

Grassberger, P., and Procaccia, I. (1983). Measuring the strangeness of strange attractors, *Physica*, **9D**, 189–208.

(1984). Dimensions and entropies of strange attractors from a fluctuating dynamics approach, *Physica*, **13D**, 34–54.

Grebogi, C., Ott, E., and Yorke, J.A. (1987). Chaos, strange attractors, and fractal basin boundaries in nonlinear dynamics, *Science*, **238**, 632 8.

Guckenheimer, J., and Holmes, P. (1983). *Nonlinear oscillations, dynamical systems, and bifurcations of vector fields*, Springer Verlag, NY.

Gutzwiller, M. (1985). Mild chaos, in *Chaotic behaviour in quantum systems*, ed. G. Casati, Plenum Press, NY, pp. 149–64.

Gwinn, E.G. and Westervelt, R.M. (1985). Intermittent chaos and low-frequency noise in the driven damped pendulum, *Phys. Rev. Lett.*, **54**, 1613–16.

(1986). Fractal basin boundaries and intermittency in the driven damped pendulum, *Phys. Rev. A.*, **33**, 4143–55.

Halsey, T., Jensen, M.H., Kadanoff, L.P., Procaccia, I., and Shraiman, B.I. (1986). Fractal measures and their singularities: the characterization of strange sets, *Phys. Rev. A*, **33**, 1141–51.

Harrison, R.G., and Biswas, D.J. (1986). Chaos in light, *Nature*, **321**, 394–401.

Hayashi, C. (1964). *Nonlinear oscillations in physical systems*, McGraw Hill, NY, p. 35.

Helleman, R.H.G. (1983). Self-generated chaotic behaviour in nonlinear systems, in *Universality in chaos*, ed. P. Cvitanovic, Adam Hilger Publ., Bristol.

Heller, E.J. (1984). Bound-state eigenfunctions of classically chaotic Hamiltonian systems: scars of periodic orbits, *Phys. Rev. Lett.*, **16**, 1515–18.

Heslot, F., Castaing, B., and Libchaber, A. (1987). Transitions to turbulence in helium gas, *Phys. Rev. A*, **43**, 5870–3.

Heutmaker, M.S., and Gollub, J.P. (1987). Wave vector field of convective flow patterns, *Phys. Rev. A*, **35**, 242–60.

Higgins, J.R. (1976). Fast Fourier transform: an introduction with some minicomputer experiments, *Am. J. Phys.*, **44**, 766–73.

Iansiti, M., Qing Hu, Westervelt, R.M., and Tinkham, M. (1985). Noise and chaos in a fractal basin boundary regime in a Josephson junction, *Phys. Rev. Lett.*, **55**, 746–9.

Jensen, R.V. (1985). Stochastic ionization of bound electrons, in *Chaotic behaviour in quantum systems*, ed. G. Casati, Plenum Press, NY, pp. 171–86.

    (1987a). Chaos in atomic physics, *Atom Phys.*, **10**, 319–32.

    (1987b). Classical chaos, *American Scientist*, **75**, 168–81.

Jensen, M.H., Bak, P., and Bohr, T. (1984). Transition to chaos by interaction of resonances in dissipative systems. I. Circle maps, *Phys. Rev. A*, **30**, 1960–9.

Kaplan, W. (1973). *Advanced calculus*, Addison–Wesley, Reading, MA, Chapter 7.

Kaplan, J.L., and Yorke, J.A. (1979). Chaotic behaviour of multi-dimensional difference equations, in *Functional differential equations and the approximation of fixed points*, eds H.O. Peitgen and H.O. Walther, Lecture notes in mathematics, **730**, Springer Verlag, Berlin, pp. 204–27.

Lorenz, E.N. (1963). Deterministic non-periodic flow, *J. Atmos. Sci.*, **20**, 130–41.

Malmstadt, H.V., Enke, C.G., and Crouch, S.R. (1981). *Electronics and instrumentation for scientists*, Benjamin/Cummings Pub. Co., Reading, MA, p. 215.

Malraison, B., Atten, P., Bergé, P., and Dubois, M. (1983). Dimension of strange attractors: an experimental determination for the chaotic regime of two chaotic systems, *J. Phys. Lett. (Paris)*, **44**, 897–902.

May, R.M. (1976). Simple mathematical models with very complicated dynamics, *Nature*, **26**, 459–67.

Moon, F.C. (1987). *Chaotic vibrations*, John Wiley and Sons, NY.

Moon, F.C., and Li, G.-X. (1985). The fractal dimension of the two-well potential strange attractor, *Physica*, **17D**, 99–108.

Orear, J. (1979). *Physics*, John Wiley and Sons, NY, Chapter 25.

Poincaré, H. (1913). *The foundation of science: science and method*, English translation, 1946, The Science Press, Lancaster, PA, p. 397.

Press, W.H., Flannery, B.P., Teukolsky, S.A., and Vetterling, W.T. (1986). *Numerical recipes: the art of scientific computing*, Cambridge Univ. Press, Cambridge.

Rapp, P.E. (1986). Oscillations and chaos in cellular metabolism and physiological systems, in *Chaos*, ed. A.V. Holden, Princeton Univ. Press, Princeton, pp. 179–208.

Robinson, H. (1921). *The mind in the making*, Harper and Brothers, NY, p. 52.

Sagdeev, R.Z., Usikov, P.A., and Zaslavsky, G.M. (1988). *Nonlinear physics: from the pendulum to turbulence and chaos*, Harwood Academic Publishers, Chur, Switzerland.

Sinai, Y.G. (1970). Dynamical systems with elastic reflections, *Russ. Math. Surv.*, **25**, 137–89.

Smale, S. (1963). Diffeomorphisms with many periodic points, in *Differential and combinatorial topology*, ed. S.S. Cairns, Princeton Univ. Press, Princeton, pp. 63–80.

Swinney, H.L. (1983). Observations of order and chaos in nonlinear systems, *Physica*, **7D**, 3–15.

Swinney, H.L., and Gollub, J.P. (1986). Characterization of hydrodynamic strange attractors, *Physica D*, **18**, 448–54.

Thomas, G.B., and Finney, R.L. (1979). *Calculus and analytic geometry*, Addison–Wesley, Reading, MA, p. 718.

Wolf, A. (1986). Quantifying chaos with Lyapunov exponents, in *Chaos*, ed. A.V. Holden, Princeton Univ. Press, Princeton, pp. 273–90.

Wolf, A., Swift, J.B., Swinney, H.L., and Vastano, J.A. (1985). Determining Lyapunov exponents from a time series, *Physica*, **16D**, 285–317.

Index

Diskette order information

A diskette that accompanies this book may be purchased direct from one of the authors. The diskette has a menu-driven runtime program — CHAOS — that accesses compiled programs which can duplicate or vary the simulations discussed in this book. The $5\frac{1}{4}$ in diskette is formatted for IBM PC, XT, and AT compatible machines with 512K memory and Hercules, CGA, or EGA graphics. CHAOS is executable without special language system diskettes, and is therefore inaccessible for modification. The True BASIC[tm] source code for the compiled programs is also provided on the diskette for those who wish to modify the programs or translate them into another language. (True BASIC is a trademark of True BASIC, Inc.)

The price of the diskette is $12. Payment should be sent in the form of *(a)* a check from US residents, or *(b)* a Postal Money Order in US funds from those living outside the United States, made payable to: Gregory Baker.

Send orders to: Gregory Baker
P.O. Box 278
Bryn Athyn, PA
USA 19009

Mailing information should be clearly *printed* on a piece of paper which will be used as an address label.